Requirements Engineering for Software and Systems

Titles in the
Auerbach Series on Applied Software Engineering
Phillip A. Laplante, Pennsylvania State University, Series Editor

Requirements Engineering for Software and Systems
Phillip A. Laplante
1-4200-6467-3

Building Software: A Practioner's Guide
Nikhilesh Krishnamurthy and Amitabh Saran
0-8493-7303-4

Global Software Development Handbook
Raghvinder Sangwan, Matthew Bass, Neel Mullick, Daniel J. Paulish,
and Juergen Kazmeier
0-8493-9384-1

Antipatterns: Identification, Refactoring, and Management
Phillip A. Laplante and Colin J. Neill
0-8493-2994-9

Software Engineering Quality Practices
Ronald Kirk Kandt
0-8493-4633-9

Requirements Engineering for Software and Systems

Phillip A. Laplante

CRC Press is an imprint of the
Taylor & Francis Group, an **informa** business
AN AUERBACH BOOK

Auerbach Publications
Taylor & Francis Group
6000 Broken Sound Parkway NW, Suite 300
Boca Raton, FL 33487-2742

Printed in the United States of America on acid-free paper
10 9 8 7 6 5 4 3 2 1

International Standard Book Number-13: 978-1-4200-6467-4 (Hardcover)

Library of Congress Cataloging-in-Publication Data

Laplante, Phillip A.
Requirements engineering for software and systems / Phillip A. Laplante.
p. cm. -- (Auerbach series on applied software engineering)
Includes bibliographical references and index.
ISBN 978-1-4200-6467-4 (hardcover : alk. paper)
1. Software engineering. 2. System design. 3. Requirements engineering. I. Title.

QA76.758.L326 2009
005.1--dc22 2009002140

Visit the Taylor & Francis Web site at
http://www.taylorandfrancis.com

and the Auerbach Web site at
http://www.auerbach-publications.com

Contents

Acknowledgments

Dr. George Hacken of the New York Metropolitan Transportation Authority contributed many ideas and inspiration used in Chapter 6, on formal methods.

Professor Larry Bernstein of Stevens Institute of Technology reviewed the first draft of the manuscript and provided many suggestions for improvement.

Dr. Colin Neill and Dr. Raghu Sangwan of Penn State read portions of the text and provided valuable feedback in many discussions.

Over the years, many students read drafts of portions of the text and provided ideas, exercises, critical feedback, and examples. The author would like to particularly mention Brad Bonkowski for contributing the prototype software requirements specification found in the appendix, and Jim Kelly, George Kniaz, Michelle Reichart, Ann Richards, Ryan Oboril, Sam Okyne, and Julie Palmer. The author would also like to thank the staff at Taylor & Francis—in particular John Wyzalek, Acquisitions Editor, and Amy Rodriguez, Project Editor. The author also wishes to thank his long-suffering wife, Nancy, and children Chris and Charlotte, for their patience and support.

I would like to thank Dr. Mark Ardis of Stevens Institute of Technology, and Dr. George Hacken of the New York Metropolitan Transit Authority for pointing out several typographical errors in the first printing, which have since been corrected.

Of course, any errors of commission or omission are due to the author alone.

Introduction

Solid requirements engineering has increasingly been recognized as the key to improved on-time and on-budget delivery of software and systems projects. Nevertheless, few undergraduate engineering programs stress the importance of this discipline. Recently, however, some software programs are introducing requirements engineering as mandatory in the curriculum. In addition, new software tools are emerging that are empowering practicing engineers to improve their requirements engineering habits. However, these tools are not usually easy to use without significant training, and many working engineers are returning for additional courses that will help them understand the requirements engineering process.

This book is intended to provide a comprehensive treatment of the theoretical and practical aspects of discovering, analyzing, modeling, validating, testing, and writing requirements for systems of all kinds, with an intentional focus on software-intensive systems. This book brings into play a variety of formal methods, social models, and modern requirements writing techniques to be useful to the practicing engineer.

Audience

This book is intended for professional software engineers, systems engineers, and senior and graduate students of software or systems engineering. Much of the material is derived from the graduate level "Requirements Engineering" course taught at Penn State's Great Valley School of Graduate and Professional Studies, where the author works. The typical student in that course has five years of work experience as a software professional and an undergraduate degree in engineering, science, or business. Typical readers of this book will have one of the following or similar job titles:

- Software engineer
- Systems engineer
- Sales engineer

- Systems analyst
- [XYZ] engineer (where "XYZ" is an adjective for most engineering disciplines, such as "electrical" or "mechanical")
- Project manager
- Business analyst
- Technical architect
- Lead architect

Exemplar Systems

Before proceeding, three systems are presented that will be used for running examples throughout the book. These systems were selected because they involve application domains with which most readers are bound to be familiar, and because they cover a wide range of applications from embedded to organic in both industrial and consumer implementations. Consider a domain, however, with which you (and this author) are likely unfamiliar, say, mining of anthracite coal. Imagining the obvious difficulties in trying to communicate about such a system highlights the importance of domain understanding in requirements engineering. This topic will be discussed at length in Chapter 1.

The first system to be introduced is an airline baggage handling system, probably similar to the system found in the bowels of every major airport and known to "eat" your baggage. Check-in clerks and baggage handlers tag your bags at check-in with a barcode ID tag. Then the baggage is placed on a conveyor belt where it moves to a central exchange point, and is redirected to the appropriate auxiliary conveyor for loading on an airplane-bound cart or a baggage carousel. Along the way, the system may conduct some kind of image scan and processing to detect the presence of unauthorized or dangerous contents (such as weapons or explosives). A baggage handling system is an embedded, real-time system; that is, the software is closely tied to the hardware, and deadline satisfaction is a key goal of the system.

The Denver International Airport tried to build a very sophisticated version of such a system several years ago. The system used PCs, thousands of remote-controlled carts, and a 21-mile-long track. Carts moved along the track, carrying luggage from check-in counters to sorting areas and then straight to the flights waiting at airport gates. After spending $230 million over 10 years, the project was cancelled.* Much of the failure can be attributed to requirements engineering mistakes.

The second exemplar system is a point of sale system for one location of a large pet store chain. This type of system would provide such capabilities as cashier functions and inventory tracking, tax reporting, and end-of-year closeout. It might handle self-checkout, coupon scanning, product returns, and more. This is

* de Neufville, R. (1994) The Baggage System at Denver: Prospects and Lessons, *Journal of Air Transport Management*, 1(4) Dec., 229–236.

a transaction-oriented business domain application, and although there are many available off-the-shelf systems, let's assume there is a need for a custom system. This type of system is organic; that is, it is not tied to any specialized hardware. PCs or PC-based cash registers, storage devices, and network support comprise the main hardware.

The final system that we will be following is for a "Smart Home," that is, a home in which one or more PCs control various aspects of the home's climate control, security, ambience, entertainment, and so forth. A Smart Home is a consumer application and a semi-detached system—it uses some specialized but off-the-shelf hardware and software. We will imagine that this Smart Home is being built for someone else (not you) so that you can remain objective about its features.

For the first two systems, we'll see many more details as we go along. For the third system, the reader will be referred to the appendix, which includes a complete (though imperfect, for illustrative purposes) example of a software requirements specification for the Smart Home. I am grateful to my former student Bradley Bonkowski for preparing this example.

For the purposes of experimentation and practices, you are encouraged to select another appropriate system to play with. Some candidate systems are

- The passenger safety restraint system for an automobile of your choice.
- A system that is intended as a "personal assistant" to all members of a household. That is, the system will store such information as phone numbers, music files, video files, calendar information, maps and directions, and so forth and will provide various alerting, reporting, "data mining," and business logic features to each user.
- A family tree maker and genealogical database system.
- A game of your choosing (keep it simple—3-D chess or checkers or popular board games are best).
- A simulation of some familiar enterprise (e.g., a traffic intersection, elevator control, manufacturing environment).
- Any aspect of your work that seems suitable to automation.

You can use your imagination, consult the many resources that are available on the Web, and have fun as you learn to "scope out" one or another of these systems.

Notes on Referencing and Errors

The author has tried to uphold the highest standards for giving credit where credit is due. Each chapter contains a list of related readings, and they should be considered the primary references for that chapter. Where direct quotes or non-obvious facts are used, an appropriate note or in-line citation is provided. In particular, it is noted where the author published portions in preliminary form in other scholarly publications.

Despite these best efforts and those of the reviewers and publisher, there are still likely errors to be found. Therefore, if you believe that you have found an error—whether it is a referencing issue, factual error, or typographical error, please contact the author at plaplante@psu.edu. If your correction is accepted, you will be acknowledged in future editions of the book.

Disclaimers

Great pains have been taken to disguise the identities of any organizations as well as individuals that are mentioned. In every case, the names and even elements of the situation have been changed to protect the innocent (and guilty). Therefore, any similarity between individuals or companies mentioned herein is purely coincidental.

About the Author

Philip A. Laplante is professor of software engineering and a member of the graduate faculty at The Pennsylvania State University. His research, teaching, and consulting focuses on software quality, particularly with respect to requirements, testing, and project management. He is also the chief technology officer of the Eastern Technology Council, a nonprofit business advocacy group serving the greater Philadelphia metropolitan area. As CTO, he created and directs the CIO Institute, a community of practice of regional CIOs. Before joining Penn State he was a professor and senior academic administrator at several other colleges and universities.

Prior to his academic career, Dr. Laplante spent nearly eight years as a software engineer and project manager working on avionics (including the Space Shuttle), CAD, and software test systems. He was also director of business development for a software consulting firm. He has authored or edited 24 books and more than 150 papers, articles, reviews, and editorials.

Dr. Laplante received his B.S., M.Eng., and Ph.D. in computer science, electrical engineering, and computer science, respectively, from Stevens Institute of Technology and an MBA from the University of Colorado. He is a fellow of the IEEE and SPIE and a member of numerous professional societies, program committees, and boards. He has consulted to Fortune 500 companies, small businesses, the U.S. DOD, and NASA on technical and management issues. He also serves as a CIO/CEO coach.

Chapter 1

Introduction to Requirements Engineering

Motivation

Very early in the drive to industrialize software development, Royce (1975) pointed out the following truths:

> There are four kinds of problems that arise when one fails to do adequate requirements analysis: top-down design is impossible; testing is impossible; the user is frozen out; management is not in control. Although these problems are lumped under various headings to simplify discussion, they are actually all variations of one theme—poor management. Good project management of software procurements is impossible without some form of explicit (validated) and governing requirements.

These truths still apply today, and a great deal of research has verified that devoting systematic effort to requirements engineering can greatly reduce the amount of rework needed later in the life of the software product and can improve various qualities of the software cost effectively.

Too often systems engineers forego sufficient requirements engineering activity either because they do not understand how to do requirements engineering properly or because there is a rush to start coding (in the case of a software product).

Clearly, these eventualities are undesirable, and it is a goal of this book to help engineers understand the correct principles and practices of requirements engineering.

What Is Requirements Engineering?

There are many ways to portray the discipline of requirements engineering depending on the viewpoint of the definer. For example, a bridge is a complex system, but has a relatively small number of patterns of design that can be used (for example, suspension, trussed, cable-stayed). Bridges also have specific conventions and applicable regulations in terms of load requirements, materials that are used, and the construction techniques employed. So when speaking with a customer (for example, the state department of transportation) about the requirements for a bridge, much of its functionality can be captured succinctly:

> The bridge shall replace the existing span across the Brandywine River at Creek Road in Chadds Ford, Pennsylvania, and shall be a cantilever bridge of steel construction. It shall support two lanes of traffic in each direction and handle a minimum capacity of 100 vehicles per hour in each direction.

Of course there is a lot of information missing from this "specification," but it substantially describes what this bridge is to do.

Other kinds of systems, such as biomechanical or nanotechnology systems with highly specialized domain language, have seemingly exotic requirements and constraints. Still other complex systems have so many kinds of behaviors that need to be embodied (even word processing software can support thousands of functions) that the specification of said systems becomes very challenging indeed.

Since the author is a software engineer, we reach out to that discipline for a convenient, more-or-less universal definition for requirements engineering that is due to Pamela Zave:

> Requirements engineering is the branch of software engineering concerned with the *real-world goals* for, functions of, and constraints on software systems. It is also concerned with the relationship of these factors to *precise specifications* of software behavior, and to their *evolution over time and across software families.* (Zave 1997)

But we wish to generalize the notion of requirements engineering to include any system, whether it be software only, hardware only, or hardware and software (and many complex systems are a combination of hardware and software), so we rewrite Zave's definition as follows:

> Requirements engineering is the branch of **engineering** concerned with the *real-world goals* for, functions of, and constraints on **systems**. It is also concerned with the relationship of these factors to *precise specifications* of **system** behavior and to their *evolution over time and across families of **related systems***.

The changes we have made to Zave's definition are in **bold**. We refer to this modified definition when we speak of "requirements engineering" throughout this text. And we will explore all the ramifications of this definition and the activities involved in great detail as we move forward.

You Probably Don't Do Enough Requirements Engineering

In 2003 the author and a colleague conducted an exploratory, Web-based survey of nearly 2000 software and systems practitioners from our region (Southeast Pennsylvania) to determine the pervasiveness of object-oriented and formal methodologies, as well as the perceived real value of these approaches. Almost 200 individuals responded, which is an unusually high number for an unsolicited survey. Our survey uncovered some surprising findings, and those related to requirements engineering will be discussed in this text. A detailed description of the survey methodology can be found in Neill and Laplante (2003) and Laplante and Neill (2004).

One of the first surprising findings related to the amount of requirements engineering perceived to be done by the survey respondents (Figure 1.1).

When the question was asked "Does your organization do enough requirements engineering?" 19% of the 157 who responded to this question didn't know. There are lots of reasons why the respondents might have such a response, so we can't draw too strong a conclusion. But it is probably the case in an organization where requirements engineering is robust that this fact would be well known throughout the organization. In any case, it is astounding that 52% of the respondents thought that their organization did not do enough requirements engineering. It is nice to

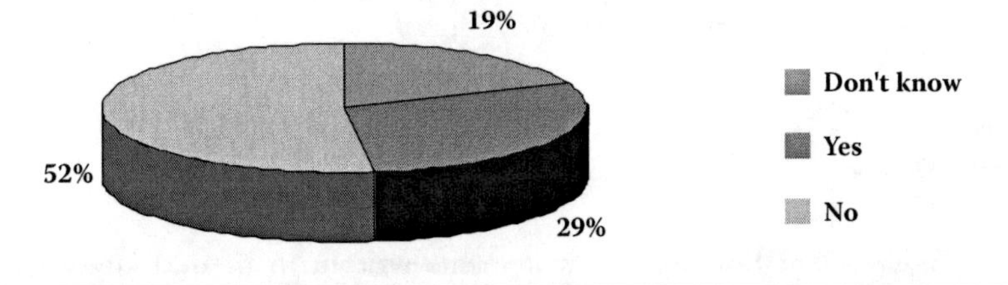

Figure 1.1 Do you do enough requirements engineering? 157 responses (Neill and Laplante 2003).

believe that we can extrapolate the results of the response to this question to include software engineers and systems engineers beyond the sample region of the survey.

What Are Requirements?

Part of the challenge in requirements engineering has to do with an understanding of what a "requirement" really is. Requirements can range from high-level, abstract statements and back-of-the-napkin sketches to formal (mathematically rigorous) specifications. These varying representation forms occur because stakeholders have needs at different levels, hence, depend on different abstraction representations. Stakeholders also have varying abilities to make and read these representations (for example a business customer versus a design engineer), leading to diverse quality in the requirements. We will discuss the nature of stakeholders and their needs and capabilities in the next chapter.

Requirements Versus Goals

A fundamental challenge for the requirements engineer is recognizing that customers often confuse requirements and goals (and engineers sometimes do too).

Goals are high-level objectives of a business, organization, or system, but a *requirement* specifies how a goal should be accomplished by a proposed system. So, to the state department of transportation, a goal might be "to build the safest bridge in the world." What is really intended, however, is to stipulate performance requirements on the bridge materials, qualifications of the contractor and engineers, and building techniques that will lead to a safe bridge.

To treat a goal as a requirement is to invite trouble because achievement of the goal will be difficult to prove. In addition, goals evolve as stakeholders change their minds and refine and operationalize goals into behavioral requirements.

Requirements Level Classification

To deal with the diversity in requirements types, Sommerville (2005) suggests organizing them into three levels of abstraction:

- User requirements
- System requirements
- Software design specifications

User requirements are abstract statements written in natural language with accompanying informal diagrams. They specify what services (user functionality) the system is expected to provide and any constraints. In many situations user stories can play the role of user requirements.

System requirements are detailed descriptions of the services and constraints. System requirements are sometimes referred to as *functional specification* or *technical annex*. These requirements are derived from analysis of the user requirements. They act as a contract between client and contractor, so they should be structured and precise. Use cases can play the role of system requirements in many situations.

Finally, software design specifications emerge from the analysis and design documentation used as the basis for implementation by developers. The software specification is essentially derived directly from analysis of the system specification. The software requirements specification document (SRS) is the "contractual" document that we generally refer to when we speak of a "software" or "system" requirements specification. In the case of a hybrid hardware/software system, the SRS would include both detailed design elements for the software as well as hardware (e.g., schematic diagrams and logic diagrams). In the case of a purely mechanical system, design drawings take the place of the SRS.

To illustrate the differences in these specification levels, consider the following from the airline baggage handling system:

A user requirement

- The system shall be able to process 20 bags per minute.

Some related system requirements

- Each bag processed shall trigger a baggage event.
- The system shall be able to handle 20 baggage events per minute.

Finally, the associated system specifications

1.2 The system shall be able to process 20 baggage events per minute in operational mode.
  1.2.1 If more than 20 baggage events occur in a one-minute interval, then the system shall …
  1.2.2 [more exception handling] …

For the pet store POS system, consider the following:

A user requirement

- The system shall accurately compute sale totals including discounts, taxes, refunds, and rebates; print an accurate receipt; and update inventory counts accordingly.

Some related system requirements

- Each sale shall be assigned a sales ID.
- Each sale may have one or more sales items.
- Each sale may have one or more rebates.
- Each sale may have only one receipt printed.

Finally, the associated software specifications

1.2 The system shall assign a unique sales ID number to each sale transaction.
    1.2.1 Each sales ID may have zero or more sales items associated with it, but each sales item must be assigned to exactly one sales ID

The systems specification in the appendix also contains numerous specifications organized by level for your inspection.

Requirements Specifications Types

Another taxonomy for requirements specifications focuses on the type of requirement from the following list of possibilities:

- Functional requirements
- Nonfunctional requirements (NFRs)
- Domain requirements

Let's look at these more closely.

Functional Requirements

Functional requirements describe the services the system should provide and how the system will react to its inputs. In addition, the functional requirements need to explicitly state certain behaviors that the system should not do (more on this later). Functional requirements can be high level and general (in which case they are user requirements in the sense that was explained previously) or they can be detailed, expressing inputs, outputs, exceptions, and so on (in which case they are the system requirements described before).

There are many forms of representation for functional requirements, from natural language (in our case, the English language), visual models, and the more rigorous formal methods. We will spend much more time discussing requirements representation in Chapter 4.

To illustrate some functional requirements, consider the following set for the baggage handling system.

1.1 The system shall handle up to 20 bags per minute.
1.4 When the system is in idle mode, the conveyor belt shall not move.
1.8 If the main power fails, the system shall shut down in an orderly fashion within 5 seconds.
1.41 If the conveyor belt motor fails, the system shall shut down the input feed mechanism within 3 seconds.

For the pet store POS system, the following might be some functional requirements:

4.1 When the operator presses the "total" button, the current sale enters the closed-out state.
  4.1.1 When a sale enters the closed-out state, a total for each non-sale item is computed as number of items times the list price of the item.
  4.1.2 When a sale enters the closed-out state, a total for each sale item is computed.

Nonfunctional Requirements

Nonfunctional requirements (NFRs) are imposed by the environment in which the system is to operate. These kinds of environments include timing constraints, quality properties, standard adherence, programming languages to be used, and so on.

Sommerville (2005) suggests the nonfunctional types depicted in Figure 1.2, and we will discuss a few of these.

Many of these NFRs are beyond the control of the requirements engineer and customer. For example, in the United States, privacy requirements are typically based on certain legislation (such as the Health Information Patient Privacy Act or HIPPA), as are safety requirements. Standards requirements are based on compliance with national or international standards. Interoperability requirements may be partially under the control of the requirements engineer and customer if the system in question is to operate in conjunction with other systems controlled by the customer (if a third party controls the other systems, then interoperability is an independent variable). Other nonfunctional requirements, such as delivery, usability, performance, and organization, are largely under the control of the customer. Whatever the case, all of the nonfunctional requirements need to be tracked by the requirements engineer, typically, using an appropriate software tool.

To illustrate some nonfunctional requirements, consider the baggage handling system. Some NFRs might include

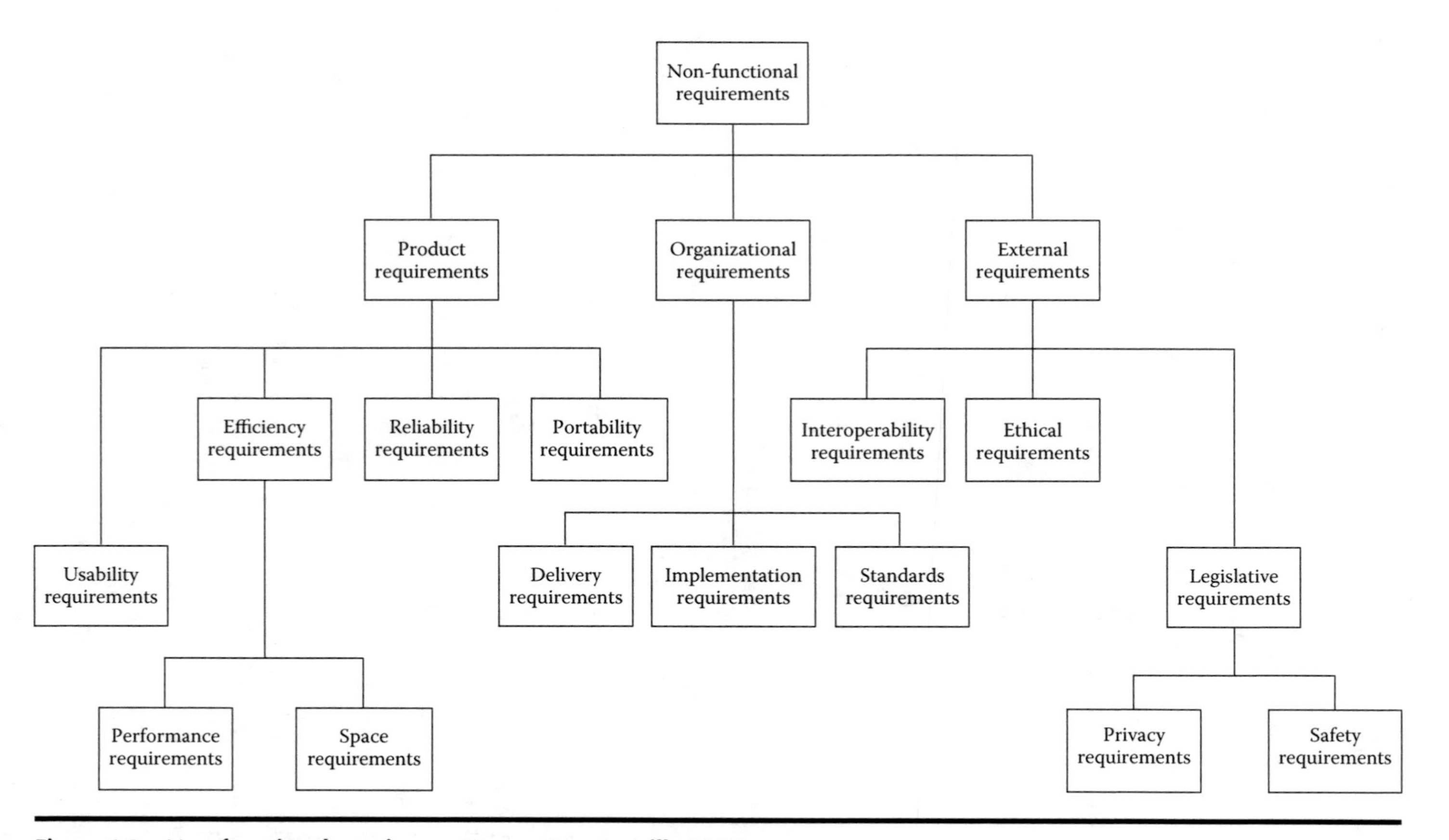

Figure 1.2 Non-functional requirement types (Sommerville 2005).

Product requirements

- Efficiency
  - Performance (e.g., number of bags per minute)
  - Space (e.g., physical size of system, amount of memory, power consumption)
- Reliability (MTBF,* MTFF†)
- Portability (e.g., can it be used with other hardware?)
- Usability (amount of training required)

Organizational requirements

- Delivery (e.g., date of delivery, date when fully operational, training sessions, updates)
- Implementation (e.g., full capability in first roll-out or phased capability)
- Standards (if there are industry standards for baggage handling systems)

External requirements

- Interoperability (e.g., with other equipment, communications standards)
- Ethical (e.g., security clearance for REs, professional certification)
- Legislative
  - Privacy (e.g., HIPPA, FERPA‡)
  - Safety (various OSHA§ regulations)

For the pet store POS system we might have these NFRs:

Product requirements

- Efficiency
  - Performance (e.g., time to finalize a sale transaction)
  - Space (e.g., physical size of system, amount of memory, power consumption)
- Reliability (MTBF, MTFF)
- Portability (e.g., can it be used with other hardware?)
- Usability (amount of training required)

Organizational requirements

- Delivery (e.g., date of delivery, fully operational, training sessions, updates)
- Implementation (e.g., full capability in first roll-out or phased?)
- Standards (if there are industry standards for POS systems)

---

* Mean time before failure.

† Mean time before first failure.

‡ Federal Educational Rights and Privacy Act (US).

§ Occupational Health and Safety Administration (US).

External requirements

- Interoperability (e.g., with other equipment, communications standards)
- Ethical (e.g., professional certification for sales associates)
- Legislative
  - Privacy (e.g., general laws governing credit card transactions)
  - Safety (various OSHA regulations)

To conclude this discussion, we note that some NFRs are difficult to define precisely, making them difficult to verify. Moreover, it is easy to confuse goals with NFRs.

Remember a goal is a general intention of a stakeholder, for example:

> The system should be easy to use by experienced operators,

whereas a verifiable NFR is a statement using some objective measure:

> Experienced operators shall be able to use the following 18 system features after two hours of hands-on instructor-led training with an error rate of no greater than 0.5%.

Here there is specificity as to which features an "experienced" user should master (though they are not listed here for brevity), a definition of the kind of training that is required (instructor led, as opposed to, say, self study), and sufficient leeway to recognize that no person is perfect.

Domain Requirements

Domain requirements are derived from the application domain. These types of requirements may consist of new functional requirements or constraints on existing functional requirements, or they may specify how particular computations must be performed.

In the baggage handling system, for example, various domain realities create requirements. There are industry standards (we wouldn't want the new system to under-perform versus other airlines' systems). There are constraints imposed by existing hardware available (e.g., conveyor systems). And there may be constraints on performance mandated by collective bargaining agreements with the baggage handlers union.

For the pet store POS system, domain requirements are imposed by the conventional store practices. For example:

- Handling of cash, credit cards, and coupons
- Display interface and receipt format

- Conventions in the pet store industry (e.g., frequent-buyer incentives, buy one get one free)
- Sale of items by the pound (e.g., horse feed) versus by item count (e.g., dog leashes)

The appendix also contains some NFRs for the Smart Home.

Domain Vocabulary Understanding

The requirements engineer must be sure to fully understand the application domain vocabulary as there can be subtle and profound differences in the use of terms in different domains. The following true incident illustrates the point. The author was once asked to provide consulting services to a very large, international package delivery company. After a few hours of communicating with the engineers of the package delivery company, it became clear that the author was using the term "truck" incorrectly. While the author believed that the "truck" referred to the familiar vehicle that would deliver packages directly to customers, the company used the term "package car" to refer to those vehicles. The term "truck" was reserved to mean any long-haul vehicle (usually an 18-wheeled truck) that carried large amounts of packages from one hub to another. So there was a huge difference in the volume of packages carried in a "truck" and a "package car." Imagine, then, if a requirement was written involving the processing of "packages from 1000 trucks" (when it was really meant "1000 package cars"). Clearly this difference in domain terminology understanding would have been significant and potentially costly.

Requirements Engineering Activities

The requirements engineer is responsible for a number of activities. These include

- requirements elicitation/discovery
- requirements analysis and reconciliation
- requirements representation/modeling
- requirements verification and validation
- requirements management

We explore each of these activities briefly.

Requirements Elicitation/Discovery

Requirements elicitation/discovery involves uncovering what the customer needs and wants. But elicitation is not like picking low-hanging fruit from a tree. While some requirements will be obvious (e.g., the POS system will have to compute sales tax), many requirements will need to be discovered/or teased from the customer through well-defined approaches. This aspect of requirements engineering also involves discovering who the stakeholders are; for example, are there any hidden stakeholders? Elicitation also involves determining the nonfunctional requirements, which are often overlooked.

Requirements Analysis and Reconciliation

Requirements analysis and reconciliation involves techniques to deal with a number of problems with requirements in their "raw" form, that is, after they have been collected from the customers. Problems with raw requirements include

- They don't always make sense.
- They often contradict one another (and not always obviously so).
- They may be inconsistent.
- They may be incomplete.
- They may be vague or just wrong.
- They may interact and are dependent on each other.

Many of the elicitation techniques that we will discuss subsequently are intended to avoid or alleviate these problems. Formal methods are also useful in this regard.

Requirements Representation and Modeling

Requirements representation and modeling involves converting the requirements processed raw requirements into some model (usual natural language, math, and visualizations). Proper representations facilitate communication of requirements and conversion into a system architecture and design. Various techniques are used for requirements representation including informal (e.g., natural language, sketches, and diagrams), formal (mathematically sound representations), and semiformal (convertible to a sound representation or is partially rigorous). Usually some combination of these is employed in requirements representation, and we will discuss these further in subsequent chapters.

Requirements Validation

Requirements validation is the process of determining if the specification is a correct representation of the customers' needs. Validation answers the question "Am I building the right product?" Requirements validation involves various semi-formal and formal methods, text-based tools, visualizations, inspections, and so on.

Requirements Management

One of the most overlooked aspects of requirements engineering, requirements management involves managing the realities of changing requirements over time. It also involves fostering traceability through appropriate aggregation and subordination of requirements and communicating changes in requirements to those who need to know.

Managers also need to learn the skills to intelligently "push back" when scope creep ensues. Using tools to track changes and maintain traceability can significantly ease the burden of requirements management. We will discuss software tools to aid in requirements engineering in Chapter 8 and requirements management overall in Chapter 9.

The Requirements Engineer

What skills should a requirements engineer have? Christensen and Chang suggest that the requirements engineer should be organized, have experience throughout the (software) engineering lifecycle, have the maturity to know when to be general and when to be specific, and be able to stand up to the customer when necessary (Christensen and Chang 1996). Christensen and Chang further suggest that the requirements engineer should be a good manager (to manage the process), a good listener, fair, a good negotiator, multidisciplinary (e.g., have a background in traditional hard sciences and engineering augmented with communications and management skills). Finally, the requirements engineer should understand the problem domain.

Gorla and Lam (2004) imply that engineers should be "thinking," "sensing," and "judging" in the Myers-Briggs sense. We could interpret this observation to mean that requirements engineers are structured and logical (thinking), focus on information gathered and do not try to interpret it (sensing), and seek closure rather than leaving things open (judging). The author's own experience confirms the observations of Christensen, Chang, Gorla, and Lam.

Requirements Engineering Paradigms

Another way to understand the nature of requirements engineering is to look at various models for the role of the requirements engineer. We have identified the following role models:

- requirements engineer as software systems engineer
- requirements engineer as subject matter expert
- requirements engineer as architect

- requirements engineer as business process expert
- the ignorant requirements engineer

There are hybrid roles for the requirements engineer from the above as well.

Requirements Engineer as Software Systems Engineer

It is likely that many requirements engineers are probably former software systems designers or developers. When this is the case, the requirements engineer can positively influence downstream development of models (e.g., the software design). The danger in this case is that the requirements engineer may begin to create a design when he should be developing requirements specifications.

Requirements Engineer as Subject Matter Expert

In many cases the customer is looking to the requirements engineer to be a subject matter expert (SME) for expertise either in helping to understand the problem domain or in understanding the customers' own wants and desires. Sometimes the requirements engineer isn't an SME—they are an expert in requirements engineering. In those cases where the requirements engineer is not an SME, consider joining forces with an SME.

Requirements Engineer as Architect

Building construction is often used as a metaphor for software construction. In our experience architects and landscape architects play similar roles to requirements engineers (and this similarity is often used as an argument that software engineers need to be licensed). Daniel Berry has written about this topic extensively (Berry 1999, 2003). Berry noted that the analogous activities reduce scope creep and better involve customers in the requirements engineering process. In addition, Zachman (1987) introduced an architectural metaphor for information systems, though his model is substantially different from the one about to be presented.

The similarities between architecture (in the form of home specification) and software/systems specification are summarized in Table 1.1. If you have been through the process of constructing or renovating a home, you will appreciate the similarities.

Requirements Engineer as Business Process Expert

The activities of requirements engineering comprise a form of problem solving—the customer has a problem and the system must solve it. Often, solving the problem at hand involves the requirements engineer advising changes in business processes to simplify expression of system behavior. While it is not the requirements engineer's role to conduct business process improvement, this side benefit is frequently realized in many cases.

Table 1.1 Architectural Model for Systems Engineering

Home Building	*Software/System Building*
Architect meets with and interviews clients. Tours property. Takes notes and pictures.	Requirements engineer meets with customers and uses interviews and other elicitation techniques.
Architect makes rough sketches (shows to clients, receives feedback).	Requirements engineer makes models of requirements to show to customers (for example, prototypes, draft SRS)
Architect makes more sketches (for example, elevations) and perhaps more sophisticated models (for example, cardboard, 3D-virtual models, fly-through animations).	Architect refines requirements and adds formal and semi-formal elements (for example, UML). More prototyping is used.
Architect prepares models with additional detail (floor plans).	Requirements engineer uses information determined above to develop complete SRS
Future models (for example, construction drawings) are for contractors' use.	Future models (for example, software design documents) are for developers' use.

Ignorance as Virtue

Berry (1995) suggested having both novices and experts in the problem domain involved in the requirements engineering process. His rationale is as follows. The "ignorant" people ask the "dumb" questions, and the experts answer these questions. Having the requirements engineer as the most "ignorant" person of the group is not necessarily a bad thing, then, at least with respect to the problem domain because it forces him to ask the hard questions and to challenge conventional beliefs. Of course, the "ignorant" requirements engineer is completely in opposition to the role of subject matter expert.

Berry also noted that using formal methods in requirements engineering is a form of ignorance because a mathematician is generally ignorant about an application domain before she starts modeling it.

Role of the Customer?

What role should the requirements engineer expect the customer to play? The customers' roles are many and include

- helping the requirements engineer understand what they need and want (elicitation and validation)
- helping the requirements engineer understand what they don't want (elicitation and validation)
- providing domain knowledge when necessary and possible
- alerting the requirements engineer quickly and clearly when they discover that they or others have made mistakes
- alerting the requirements engineer quickly when they determine that changes are necessary (really necessary)
- controlling their urges to have "aha moments" that cause major scope creep
- sticking to all agreements

In particular, the customer is responsible for answering the following four questions, with the requirements engineer's help, of course:

1. Is the system that I want feasible?
2. If so, how much will it cost?
3. How long will it take to build?
4. What is the plan for building and delivering the system to me?

The requirements engineer must manage customers' expectations with respect to these issues. We will explore the nature and role of customers and stakeholders in the next chapter.

Problems with Traditional Requirements Engineering

Traditional requirements engineering approaches suffer from a number of problems, many of which we have already discussed (or will discuss) and many of which are not easily resolved, though we will discuss some of their resolutions later. These problems include

- natural language problems (e.g., ambiguity, imprecision)
- domain understanding
- dealing with complexity (especially temporal behavior)
- difficulties in enveloping system behavior
- incompleteness (missing functionality)
- over-completeness (gold plating)
- overextension (dangerous "all")
- inconsistency
- incorrectness
- and more

Natural language problems result from the ambiguities and context sensitivity of natural (human) languages. We know these language problems exist for everyone, not just requirements engineers, and lawyers and legislators make their living finding, exploiting, or closing the "loopholes" found in any laws and contracts written in natural language.

We have already discussed the issue of domain understanding and have observed, as have others, that the requirements engineer may be an expert in the domain in which the system to be built will operate. System complexity is a pervasive problem that faces all systems engineers, and this will be discussed shortly. The problems of fully specifying system behavior and of missing behavior are also very challenging, but there are some techniques that can be helpful in missing, at least, the most obvious functionality in a system.

Complexity

One of the greatest difficulties in dealing with the requirements engineering activities, particularly elicitation and representation, for most systems is that they are "complex." Without trying to invent a definition for "complexity," we contend that the challenges and complexity of capturing non-trivial behavior of any kind illustrate the notion of complex. Such difficulties are found in even the simplest of "repeatable" human endeavors.

Imagine, for example, someone asked you to describe the first five minutes of your morning from the moment you wake up. Could you do it with precision (try it)? No, you could not. Why? There are too many possible different paths your activities can take and too much uncertainty. Even with an atomic clock, you can't claim to wake up the same time every day (because even atomic clocks are imperfect). But of course, you wake up differently depending on the day of the week, whether you are on vacation from work, if it is a holiday, and so on. You have to account for that in your description. But what if you wake up sick? How does that change the sequence of events? What if you accidentally knock over the glass of water on your nightstand as you get up? Does that change the specification of the activity? Or if you trip on your dog as you head to the bathroom? We could go on with this example and repeat this exercise with other simple tasks such as mowing the lawn or shopping for food. In fact, until you constrain the problem to the point of ridiculousness, you will find it challenging or even impossible to precisely capture any non-trivial human activities.

Now consider a complex information or embedded processing system. Such a system will likely have to have interactions with humans. Even in those systems that do not directly depend on human interaction, it is the intricacy of temporal behavior as well as the problems of unanticipated events that make requirements elicitation and specification so difficult (and all the other requirements activities hard too).

Rittel and Webber (1973) defined a class of complex problems that they called "wicked." Wicked problems have ten characteristics:

- There is no definitive formulation of a wicked problem.
- Wicked problems have no stopping rule.
- Solutions to wicked problems are not true-or-false but good-or-bad.
- There is no immediate and no ultimate test of a solution to a wicked problem.
- Every solution to a wicked problem is a "one-shot operation;" because there is no opportunity to learn by trial-and-error, every attempt counts significantly.
- Wicked problems do not have an enumerable (or an exhaustively describable) set of potential solutions, nor is there a well-described set of permissible operations that may be incorporated into the plan.
- Every wicked problem is essentially unique.
- Every wicked problem can be considered a symptom of another problem.
- The existence of a discrepancy representing a wicked problem can be explained in numerous ways. The choice of explanation determines the nature of the problem's resolution
- The planner (designer) has no right to be wrong.

Rittel and Webber meant wicked problems to be of an economic, political, and societal nature (e.g., hunger, drug abuse, conflict in the Middle East); therefore, they offer no appropriate solution strategy for requirements engineering. Nevertheless, it is helpful to view requirements engineering as a wicked problem because it helps explain why the task is so difficult—because in many cases, real systems embody most if not all of the characteristics of a wicked problem.

Four Dark Corners (Zave and Jackson)

Many of the problems with traditional requirements engineering arise from "four dark corners" (Zave and Jackson 1997). We repeat the salient points here, verbatim, with commentary in italics.

1. All the terminology used in requirements engineering should be grounded in the reality of the environment for which a machine is to be built.
2. It is not necessary or desirable to describe (however abstractly) the machine to be built.
   - Rather, the environment is described in two ways as it would be without or in spite of the machine and as we hope it will become because of the machine.
   - *Specifications are the **what** to be achieved by the system, not the **how**.*
3. Assuming that formal descriptions focus on actions, it is essential to identify which actions are controlled by the environment, which actions are controlled

by the machine, and which actions of the environment are shared with the machine.

- All types of actions are relevant to requirements engineering, and they might need to be described or constrained formally.
- If formal descriptions focus on states, then the same basic principles apply in a slightly different form.
- The *method of formal representation should follow the underlying organization of the system. For example, a state-based system is best represented by a state-based formalization.*

4. The primary role of domain knowledge in requirements engineering is in supporting refinement of requirements to implementable specifications.
   - Correct specifications, in conjunction with appropriate domain knowledge, imply the satisfaction of the requirements.
   - *Failure to recognize the role of domain knowledge can lead to unfilled requirements and forbidden behavior.*

Difficulties in Enveloping System Behavior

Imagine the arbitrary system shown in Figure 1.3 with *n* inputs and *m* outputs.

The set of inputs, I, derives from human operators, sensors, storage devices, other systems, and so on. The outputs, O, pertain to display devices, actuators, storage devices, other systems, and so on. The only constraint we will place on the system is that the inputs and outputs are digitally represented within the system (if they are from/to analog devices or systems, an appropriate conversion is needed). We define a behavior of the system as an input/output pair. Since the inputs and outputs are discrete, this system can be thought of as having an infinite but countable number of behaviors, $B \subseteq I \times O$.

Imagine the behavior space, *B*, is represented by the Venn diagram of Figure 1.4. The leftmost circle in the diagram represents the desired behavior set, as it is understood by the customer. The area outside this circle is unwanted behavior and desirable behavior that, for whatever reason, the customer has not discovered.

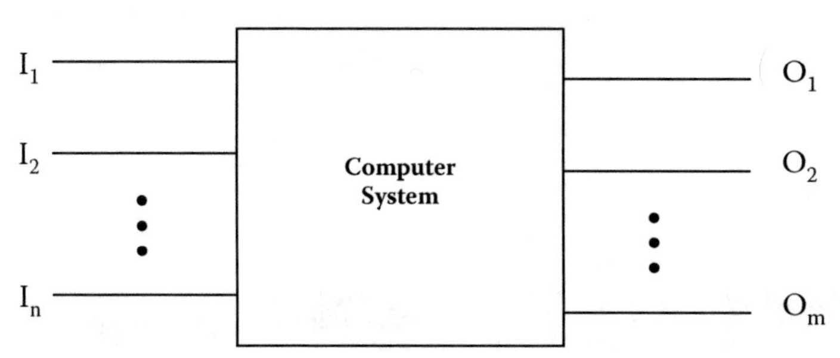

Figure 1.3 Arbitrary system with *n* inputs and *m* outputs.

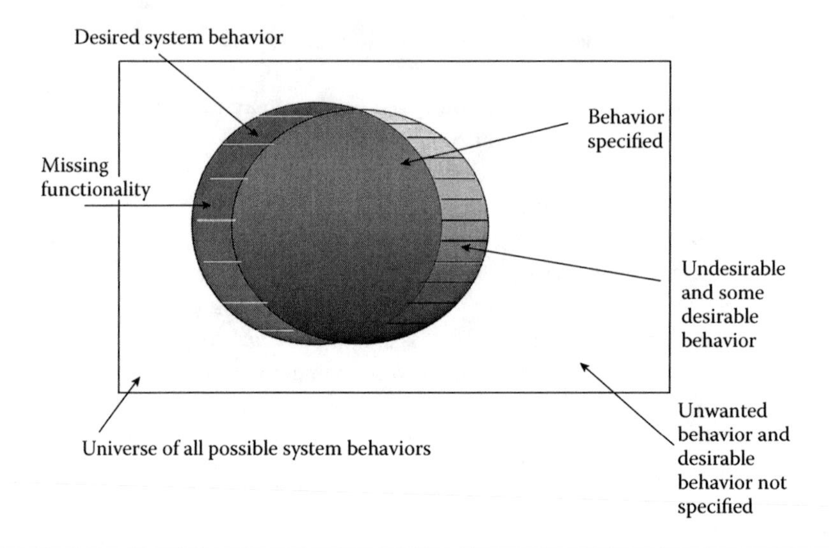

Figure 1.4 Universe of all possible behaviors, desired behavior and specified behavior.

The requirements engineer goes about his business and produces a specification that is intended to be representative of the behavior desired by the customer (the rightmost circle of Figure 1.4). Being imperfect, this specification captures some (but not all) of the desired behavior, and it also captures some undesirable behavior. The desired behavior not captured in the specified behavior is the missing functionality. The undesirable behavior that is captured is forbidden functionality.

The goal of the requirements engineer, in summary, is to make the left and right circles overlap as much as possible and to discover those missing desirable behaviors that are not initially encompassed by the specification (leftmost circle).

Recognizing that there is an infinite number of behaviors that could be excluded from the desired behaviors of the system, we need to, nevertheless, focus on the very worst of these. Therefore, some of the requirements that we will be discovering and communicating will take the form

The "system shall not…"

These kinds of behaviors are sometimes called "hazards" or "forbidden behavior." The term hazard usually refers to forbidden behavior that can lead to substantial or catastrophic failure, whereas "forbidden behavior" has a less ominous connotation.

Known "shall nots" are not a problem, but what about unknown forbidden behavior? These unknown forbidden behaviors are related to unknown risks or uncertainties in the same way that the known forbidden behaviors are related to known risks. The following tragic vignette acts as an illustration of the difference between these two concepts.

On a snowy January 13th in 1982 an Air Florida Boeing 737 jet departed from Washington DC's National Airport and crashed into the 14th Street Bridge less than a mile from the airport. Of the 79 people on board the plane, only five survived. The plane also struck seven vehicles on the bridge, killing four motorists and injuring another four.

The unfortunate passengers on the plane should have known that there was a risk of the plane crashing, no matter how small that risk may have been. In some cases, the passengers may have even taken out special crash insurance. But what were the expectations of the motorists in the vehicles on the bridge who were killed or injured? Surely, the thought of an airplane crashing into the bridge never crossed their mind. For the motorists, therefore, this crash represented an unknown but manifested uncertainty.

As an example of a forbidden behavior, consider the following for the baggage handling system:

> When the "conveyor jam" signal is set to the high state, the feed mechanism shall not permit additional items to enter the conveyor system.

Similar forbidden behaviors can be generated for the POS system.

The Danger of "All" in Specifications

Requirement specification sentences usually involve some universal quantification (e.g., "all users shall be able to access …"). But the use of "all" specifications is dangerous because they are usually not true (Berry and Kamsties 2000), so requirements involving the word "all" should be challenged and relaxed when possible.

Related to "all" specifications are "never," "always," "none," and "each" (because they can be formally equated to universal quantification), and these should be avoided as well. Finally, be careful when using the above words or their mathematical equivalents in specifications.

Exercises

1.1 What are some of the major objections and deterrents to proper requirements engineering activities?
1.2 Is requirements engineering necessary for "small" systems? Why or why not?
1.3 What are factors that may cause customers to alter requirements?
1.4 What issues may possibly arise when a requirements engineer, who is not a subject matter expert, enlists a subject matter expert to assist in defining requirements.
1.5 List some representative user requirements, system requirements, and software specifications for the pet store POS system.
1.6 List five typical functional requirements for the baggage handling system.

1.7 List five forbidden functional requirements for the pet store POS system.
1.8 Conduct some Web research to discover if there are any regulations or standards (NFR) for Smart Home systems.
1.9 For the Smart Home system what are the "hazards?" That is, make a list of what this system shall not do based on these regulations and any other information you have.

References

Berry, D. (1995) The importance of ignorance in requirements engineering, *Journal of Systems and Software*, 28:1, 179–184.

Berry, D.M. (1999) Software and house requirements engineering: Lessons learned in combating requirements creep, *Requirements Engineering Journal*, 3(3&4): 242–244.

Berry, D.M. (2003) More requirements engineering adventures with building contractors, *Requirements Engineering Journal*, 8(2):142–146.

Berry, D.M., and E. Kamsties (2000) The Dangerous 'All' in Specifications, Proceedings of the Tenth International Workshop on Software Specification and Design (IWSSD'00), San Diego, CA, 5–7 November.

Christensen, M., and C. Chang (1996) Blueprint for the ideal requirements engineer, *Software*, March, p. 12.

Gorla, N., and Yan Wah Lam (2004) Who should work with whom?: Building effective software project teams, *Communications of the ACM*, 47(6): 79–82.

Laplante, P.A., and C. Neill (2004) The demise of the waterfall model is imminent and other urban myths of software engineering, *ACM Queue*, 1(10): 10–15.

Neill, C.J., and P.A. Laplante (2003) Requirements engineering: The state of the practice, *Software*, 20(6): 40–45.

Rittel, H., and M. Webber (1973) Dilemmas in a general theory of planning, *Policy Sciences*, 4:155–169.

Royce, W. (1975) *Practical Strategies for Developing Large Software Systems*, Boston, MA: Addison-Wesley, p. 59.

Sommerville, I. (2005) *Software Engineering*, 7th edn., Boston, MA: Addison-Wesley.

Voas, J. (1999) Software hazard mining, IEEE Symposium on Application-Specific Systems and Software Engineering and Technology, pp. 180–184.

Zachman, J.A. (1987) A framework for information systems architecture, *IBM Systems Journal*, 26(3): 276–292.

Zave, P. (1997) Classification of research efforts in requirements engineering, *ACM Computing Surveys*, 29(4): 315–321.

Zave, P., and M. Jackson (1997) Four dark corners of requirements engineering, *ACM Trans. Softw. Eng. Methodology* 6(1): 1–30.

Chapter 2

Mission Statement, Customers, and Stakeholders

Mission Statements

The first thing we need to do when undertaking the development of a new system, or redesign of an old one, is to prepare a mission statement. The mission statement acts as a focal point for all involved in the system, and it allows us to weigh the importance of various features by asking the question "how does that functionality serve the mission?" In agile methodologies, to be discussed later, we would say that the mission statement plays the role of "system metaphor."

Writing mission statements can be contentious business, and many people resent or fear doing so because there can be a tendency to get bogged down on minutiae. Sometimes, mission statements tend to get very long and, in fact, evolve into a "vision" statement. A mission statement should be very short, descriptive, compelling, and never detailed, whereas a vision statement can be long. The mission statement is the "how" and the statement is the "what." A mission statement is almost a slogan.

One of the most widely cited "good" mission statements is the one associated with the Starship Enterprise from the original Star Trek series. That mission statement read

> To seek out new life, to boldly go where no man has gone before.

This statement is clear, compelling, and inspiring. And it is "useful"—fans of this classic series will recall several episodes in which certain actions to be taken by the starship crew were weighed against the mission statement.

To illustrate further, there is an apocryphal story of a multi-billion dollar-company that manufactured cardboard boxes and other packaging. The mission statement was something like

> Our mission is to provide innovative, safe, and reliable packaging to the world's manufacturers.

At a corporate retreat the senior executives began to bog down in arguments over new plants, corporate jets, marketing, and so on. In the back of the room, the CEO, a direct descendent of one of the company's founders was silently gesturing with his hands. Finally the executives noticed the CEO's actions, silenced, and asked what he was doing. He said, "I am making the shape of a box—remember what we do—we make boxes." This reference to the mission statement brought the other executives back on track.

So what might a mission statement for the baggage handling system look like? How about

> To automate all aspects of baggage handling from passenger origin to destination.

For the pet store POS system, consider

> To automate all aspects of customer purchase interaction and inventory control.

These are not necessarily clever, or awe inspiring mission statements, but they do convey the essence of the system. And they might be useful downstream when we need to calibrate the expectations of those involved in its specification. In globally distributed development in particular, the need for a system metaphor is of paramount importance.

Encounter with a Customer?

Suppose your wife (or substitute "husband," "friend," "roommate," or whoever) asks you to go to the store to pick up the following items because she wants to make a cake:

- 5 pounds flour
- 12 large eggs
- 5 pounds sugar
- 1 pound butter

Off you go to the nearest convenience mart (which is close to your home). At the store, you realize that you are not sure if she wants white or brown sugar. So you call her from your cell phone and ask which kind of sugar she wants; you learn she needs brown sugar. You make your purchases and return home.

But your wife is unhappy with your selections. You bought the wrong kind of flour; she informs you that she wanted white and you bought wheat. You bought the wrong kind of butter; she wanted unsalted. You brought the wrong kind of sugar too, dark brown; she wanted light brown. Now you are in trouble.

So you go back to the mart and return the flour, sugar, and butter. You find the white flour and brown sugar, but you could only find the unsalted butter in a tub (not sticks), but you assume a tub is acceptable to her. You make your purchase and return with the items. But now you discover that you made new mistakes. The light brown sugar purchase is fine, but the white flour you brought back is bleached; she wanted unbleached. And the butter in the tub is unacceptable—she points out that unsalted butter can be found in stick form. She is now very angry with you for your ignorance.

So, you go back to the store and sheepishly return the items, and pick up their proper substitutes. To placate your wife's anger, you decide to also buy some of her favorite chocolate candy.

You return home and she is still unhappy. While you finally got the butter, sugar, and flour right, now your wife remembers that she is making omelets for supper, and that a dozen eggs won't be enough for the omelets and the cake—she needs 18 eggs. She is also not pleased with the chocolate—she informs you she is on a diet and that she doesn't need the temptation of chocolate lying around.

One more time you visit the mart and return the chocolate and the dozen eggs. You pick up 18 eggs and return home.

You think you have gotten the shopping right when she queries: "where did you buy these things?" When you note that you bought the items at the convenience mart, she is livid—she feels the prices there are too high—you should have gone to the supermarket a few miles further down the road.

We could go on and on with this example—each time your wife discovering a new flaw in your purchases, changing her mind about quantity or brand, adding new items, subtracting others, etcetera.

But what does this situation have to do with requirements engineering and stakeholders? The situation illustrates many points about requirements engineering. First, you need to understand the application domain. In this case, having a knowledge of baking would have informed you ahead of time that there are different kinds of butter, flour, and sugar and you probably would have asked focusing questions before embarking on your shopping trip. Another point from this scenario—customers don't always know what they want—your wife didn't realize that she needed more eggs until after you made three trips to the store. And there is yet one more lesson in the story: never make assumptions about what customers want—you thought that the tub butter was acceptable; it wasn't. You finally learned that even providing customers with more than they ask for (in this case her favorite chocolate) can sometimes be the wrong thing to do.

But in the larger sense, the most important lesson to be learned from this encounter with a customer is that they can be trouble. They don't always know what they want, and, even when they do, they may communicate their wishes ineffectively. Customers can change their mind and they may have high expectations about what you know and what you will provide.

Because stakeholder interaction is so important, we are going to devote this chapter to understanding the nature of stakeholders, and more specifically the stakeholders for whom the system is being built—the customers.

Stakeholders

Stakeholders represent a broad class of individuals who have some interest (a stake) in the success (or failure) of the system in question. For any system, there are many types of stakeholders, both obvious and sublime. The most obvious stakeholder of a system is the user.

We define the *user* as the class (consisting of one or more persons) who will use the system. The customer is the class (consisting of one or more persons) who is commissioning the construction of the system. Sometimes the *customer* is called the *client* (usually in the case of software systems) or *sponsor* (in the case where the system is being built not for sale, but for internal use). But in many cases the terms "customer," "client," and "sponsor" are used interchangeably depending on the context. Note that the sponsor and customer can be the same person. And often there is confusion between who the client is and who the sponsor is that can lead to many problems.

In any case, clients, customers, users, sponsors—however you wish to redefine these terms— are all stakeholders because they have a vested interest in the system. But there are more stakeholders than these. It is said that "the customer is always right, but there are more persons/entities with an interest in the system." In fact, there are many who have a stake in any new system. For example typical stakeholders for any system might include

- customers (clients, users)
- the customers' customers (in the case of a system that will be used by third parties)
- sponsors (those who have commissioned and/or will pay for the system)
- all responsible engineering and technical persons (e.g., systems, development, test, maintenance)
- regulators (typically, government agencies at various levels)
- third parties who have an interest in the system but no direct regulatory authority (e.g., standards organizations, user groups)

- society (is the system safe?)
- environment (for physical systems)

And of course, this is an incomplete list. For example, we could go on with a chain of customers' customers' customer ..., where the delivered system is augmented by a third party, augmented again, delivered to a fourth party, and so on. In any case, when we use the term "stakeholder," we need to remember that others, not just the customer are involved.

Negative Stakeholders

Negative stakeholders are those who will be adversely affected by the system. These include competitors, investors (potentially), and people whose jobs will be changed, negatively affected, or displaced by the system. There are also internal negative stakeholders—other departments who will take on more workload, jealous rivals, skeptical managers, and more. These internal negative stakeholders can provide passive-aggressive resistance and create political nightmares for all involved. All negative stakeholders have to be recognized and accounted for as much as possible.

Finally, there are always individuals who are not directly affected by systems who are, nonetheless, interested (usually opposed) to those systems, and because they may wield some power or influence, they must be considered. These interested parties include environmentalists, animal activists, religious zealots, advocates of all types, the self-interested, and so on. Some people call these kinds of individuals "gadflies," and they shouldn't be ignored.

Stakeholder Identification

It is very important to accurately and completely identify all possible stakeholders (positive and negative) for any system. Stakeholder identification is the first step the requirements engineer must take after the mission statement has been written. Why do we start with stakeholder identification? Imagine leaving out a key stakeholder—and discovering them later? Or worse, they discover that a system is being built, in which they have an interest, and they have been ignored. These Johnny-come-lately stakeholders can try to impose all kinds of constraints and requirements changes to the system that can be very costly.

Stakeholder Questions

One way to help identify stakeholders is by answering the following set of questions:

- Who is paying for the system?
- Who is going to use the system?
- Who is going to judge the fitness of the system for use?

- What agencies (government) and entities (non-government) regulate any aspect of the system?
- What laws govern the construction, deployment, and operation of the system?
- Who is involved in any aspect of the specification, design, construction, testing, maintenance, and retirement of the system?
- Who will be negatively affected if the system is built?
- Who else cares if this system exists or doesn't exist?
- Who have we left out?

Let's try this set of questions on the airline baggage handling system. These answers are not necessarily complete—over time, new stakeholders may be revealed. But by answering these questions as completely as we can now, we reduce the chances of overlooking a very important stakeholder late in the process.

- Who is paying for the system?—Airline, grants, passengers, your tax dollars.
- Who is going to use the system?—Airline personnel, maintenance personnel, travelers (at the end).
- Who is going to judge the fitness of the system for use?—Airline, customers, unions, FAA, OSHA, the press, independent rating agencies.
- What agencies (government) and entities (non-government) regulate any aspect of the system?—FAA, OSHA, union contracts, state and local codes.
- What laws govern the construction, deployment, and operation of the system?—Various state and local building codes, federal regulations for baggage handling systems, OSHA laws.
- Who is involved in any aspect of the specification, design, construction, testing, maintenance, and retirement of the system?—Various engineers, technicians, baggage handlers union, etc.
- Who will be negatively affected if the system is built?—Passengers, union personnel.
- Who else cares if this system exists or doesn't exist?—Limousine drivers.
- Who have we left out?

And let's try this set of questions on the pet store POS system.

- Who is paying for the system?—Pet store, consumers.
- Who is going to use the system?—Cashiers, managers, customers (maybe if self-service provided). Who else?
- Who is going to judge the fitness of the system for use?—Company execs, managers, cashiers, customers. Who else?
- What agencies (government) and entities (non-government) regulate any aspect of the system?—Tax authorities, governing business entities, pet store organizations, better business bureau. What else?

- What laws govern the construction, deployment, and operation of the system?—Tax laws, business, and trade laws? What else?
- Who is involved in any aspect of the specification, design, construction, testing, maintenance, and retirement of the system?—Various engineers, CFO, managers, cashiers. We need to know them all.
- Who will be negatively affected if the system is built?—Manual cash register makers, inventory clerks. Who else?
- Who else cares if this system exists or doesn't exist?—Competitors, vendors of pet products. Who else?
- Who have we left out?

Stakeholder/Customer Classes

Assuming that we have answered the stakeholder identification questions as completely as we can, the next step is to cluster these stakeholders into classes. Because many of the stakeholders we have identified are likely to be customers, and because the other stakeholders that we have identified are probably not going to be directly involved in the requirements elicitation process (we are not going to interview legislators, for example), we cluster as many stakeholders as possible into user classes.

The reason that we cluster the classes is that for economy we need to identify a small number of individuals with whom to have contact in order to represent the interests and wants of a large swath of customers/stakeholders. So, for each of the classes that we identify we'll select an individual or very small representative group for that class and deal only with those "champions." As for the other stakeholders, we will remain wary of their concerns (and legal implications) throughout the process, but we are not likely to consult with any representatives of those classes of stakeholder if we can avoid it.

We cluster customers into classes according to interests, scope, authorization, or other discriminating factors (in effect, these are equivalence classes). Then we select a champion or representative group for each user class. Then we select the appropriate technique(s) to solicit initial inputs from each class. This is the elicitation process that will be discussed in the next chapter.

To illustrate the clustering of stakeholders/users into representative classes consider the baggage handling system. We might select the following representative classes:

- System maintenance personnel (who will be making upgrades and fixes)
- Baggage handlers (who interact with the system by turning it on and off, increasing/decreasing speed, and so on)
- Airline schedulers/dispatchers (who assign flights to baggage claim areas)
- Airport personnel (who reassign flights to different baggage claim areas)
- Airport managers and policy makers

For the pet store POS system we might have the following classes:

- Cashiers
- Managers
- System maintenance personnel (to make upgrades and fixes)
- Store customers
- Inventory/warehouse personnel (to enter inventory data)
- Accountants (to enter tax information)
- Sales department (to enter pricing and discounting information)
- Others?

For each of these classes, we'll want to select a champion and deal with them during the elicitation process. More on this aspect in the next chapter.

Customer Wants and Needs

We mentioned that the primary goal of the requirements engineer is to understand what the customers want. You also might think that you must give the customer what he needs, but remember the admonition about substituting your own value system for someone else's—what you think the customer needs might be something that they don't want. That being said, it is always helpful to reveal new functionality to the customers so that they can determine if there are features that they need but had not considered. This situation is especially relevant when the customer has less domain knowledge than the requirements engineer. At some point, however, you will need to reconcile wants and needs—requirements prioritization will be very useful here.

What Do Customers Want?

So, the objective is to satisfy customer wants. But it is not always easy to know what their wants are. Why? Because their wants exist on many levels—practical needs (e.g., minimum functionality of the system), competitive needs (it should be better than the competitors' systems), selfish needs (they want to show off and tout the system's features), and more. And sometimes the customers want "it all" and they don't want to pay a lot for it. Requirements engineers have to help them understand this reality.

One way to understand the needs levels of customers is to revisit Maslow's classic hierarchy of self-actualization (Figure 2.1).

Maslow theorized that human beings will seek to satisfy (actualize) their needs starting with the most basic (the bottom of the pyramid) and work their way up to the most elusive, esoteric, and spiritual. But they will never move very far up the pyramid (or stay at a higher level) if lower levels of needs/wants are not satisfied first.

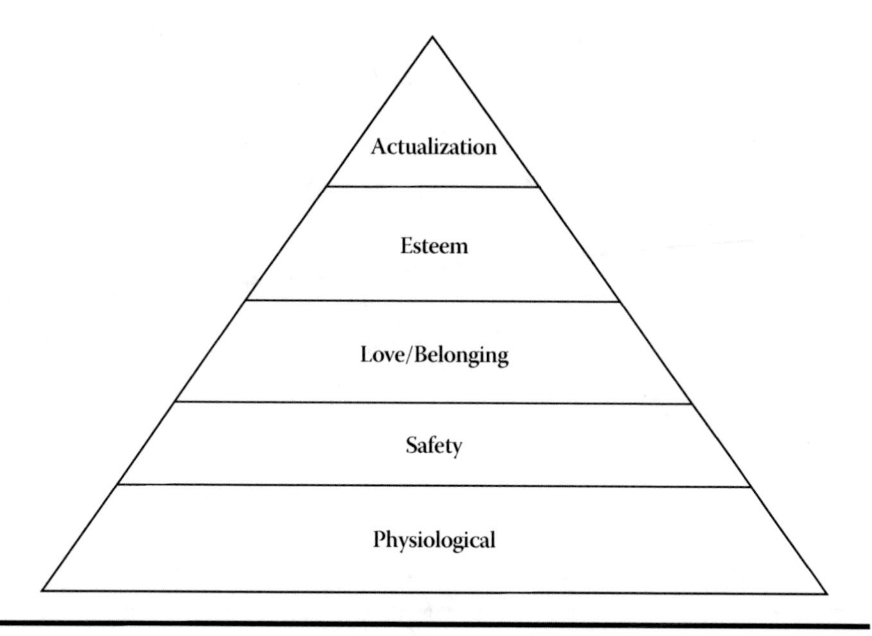

Figure 2.1 Maslow's hierarchy.

Basic needs include food, water, and sleep. These take precedence over one's physical safety, however—if you were starving, you would risk crossing a very busy street to get food on the other side. People risk jail time by stealing bread. Higher up the Maslow's pyramid is the need to be loved and to belong to some group, but he presumes that these needs are fundamentally subordinated to the need for physical safety. You might argue about this one, but the thinking is that some people will sacrifice the chance for love in order to preserve their physical wellbeing (would you continue belonging to the Sky Diving Club just because you liked one of its members?). Next, one's self-esteem is important, but not as important as the need to belong and be loved (which is why you will humiliate yourself and dress in a Roman costume for your crazy sister-in-law's wedding). Finally, Maslow defined self-actualization as "man's desire for fulfillment, namely to the tendency for him to become actually in what he is potentially: to become everything that one is capable of becoming ..." (Maslow 1943).

A variation of Maslow's hierarchy, depicted in Figure 2.2, can help explain the needs and wants of customers. Here the lowest level is basic functionality. Being a point of sale system implies certain functions must be present—such as create a sale, return an item, update inventory, and so on. At the enabling level, the customer desires features that provide enabling capabilities with respect to other systems (software, hardware, or process) within the organization. So the POS system ties into some management software that allows for real-time sales data to be tracked by managers for forecasting or inventory control purposes. The functionality at the enabling level may not meet or exceed competitors' capabilities. Those functional

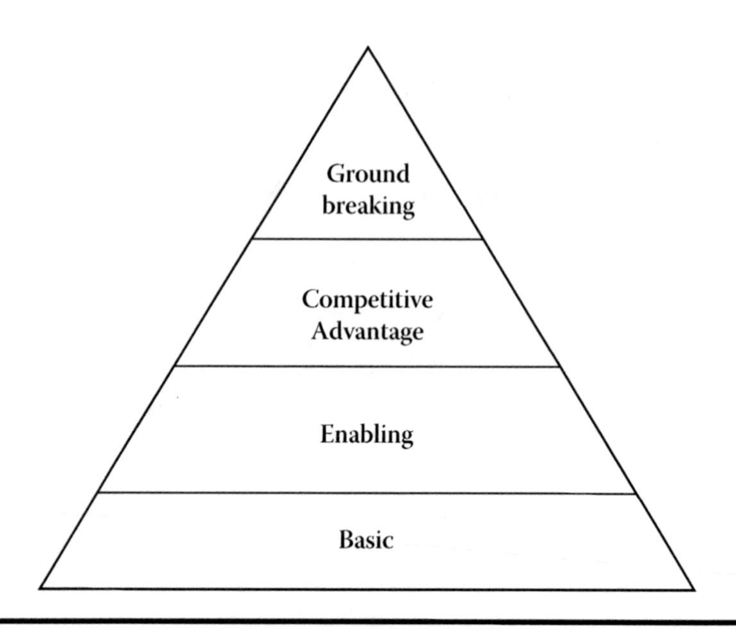

Figure 2.2 Hierarchy of customer's needs/wants.

needs are met at the competitive advantage level. Here the customer wishes for this new system to provide capabilities that exceed those of the competition or otherwise create a business advantage. Finally, ground-breaking desires would imply development of technology that exceeds current theory or practice and has implications and applications beyond the system in question. For example, some kind of new data-mining technology might be desired that exceeds current technologies. As with the Maslow hierarchy, the idea is that the lower-level functionality must not be sacrificed to meet the higher levels of functionality.

While this hierarchy implies four levels of importance of need, it is likely that in any situation there can be more or fewer levels. But the basic idea is to discover and organize customers' needs according to some meaningful hierarchy, which will be most helpful with requirements prioritization later. And this is not the first time that Maslow's theory was used to explain the needs of customers or users. For example, Valacich et al. described a modified four-level Maslow hierarchy to explain user preferences in Web-based user interfaces (Valacich et al. 2007).

In any case, let's return to our examples to consider some of the wants for the baggage handling and pet store POS systems.

For an airline baggage handling system customers probably want

- safety
- speed
- reliability
- fault-tolerance (no broken luggage!)

- maintainability
- and so on

For the pet store POS system, customers want

- speed
- accuracy
- clarity (in the printed receipt)
- efficiency
- ease of use (especially if self-service provided)
- and more

So we would use our best efforts to attend to these needs. The problem becomes, however, how do we measure satisfaction of these needs? Because, if these wants and desires cannot be measured, then we will never achieve them. We will discuss the issue of requirements satisfaction in Chapter 5.

What Don't Customers Want?

Sometimes customers are very explicit in what they don't want the system to do. These specific undesirable features or "do not wants" are frequently overlooked by the requirements engineer. Unwanted features can include

- undesirable performance characteristics
- esthetic features
- gold plating (excessive and unnecessary features)
- safety concerns ("hazards")

The "shall not" requirements are often the hardest to capture. Sometimes customers don't know what they don't want until they see it! For example, upon seeing the delivered system (or a prototype), they exclaim:

> "I know I said I wanted it to do that, but I guess I really didn't mean that."

For illustrative purposes, here are some examples of unwanted features of the baggage handling system:

- The system shall not shut down if main airport power is lost.
- The system shall not cause a failure in the main airline computer system.
- The system shall not cause baggage to be destroyed at a rate higher than 1 bag per minute.

You can see how hard it is to describe what the system is not supposed to do. More on this later.

And here are some for the pet store POS system:

- If the register tape runs out, the system shall not crash.
- If a product code is not found, the system shall not crash.
- If a problem is found in the inventory reconciliation code, the current transaction shall not be aborted.

Many of the elicitation techniques that we will discuss in the next chapter will tend to uncover unwanted features, but the requirements engineer should always try to discover what the customer does not want as much as what the customer does want.

Why Do Customers Change Their Minds?

One of the greatest challenges in dealing with customers is that they sometimes don't know precisely what they want the system to do. Why is this so? First, no one is omnipotent. A customer can't see every possible desideratum, unless they want a complete replica of another system. Helping the customer find these new features as early as possible is the job of the requirements engineer.

Another reason why customers change their minds is that, with varying levels of requirements (e.g., features, constraints, business rules, quality attributes, etc.), what is important might change as these "non-requirements" (e.g., business rules) change during the system life cycle.

Sometimes the environment in which the system functions and the customers operate changes (physical changes, economic changes, competition, regulatory changes, etc.). For example, while building the pet store POS system, state taxation rules about pet toys might change (maybe only squeaky toys are taxed), and this event might require a significant redesign of the taxation portion of the code.

Some requirements are so "obvious" that the customer doesn't think to stipulate them. For example, in what order do you print a list of customers in the pet store POS system? The customer's "obvious" answer is alphabetically by last name, but the system designer thought numerically by customer id.

Another reason for changing requirements has to do with the discovery of the return on investment. In one project, for example, it was discovered that what the customer deemed the most important requirement (paperwork reduction) did not add substantial economic value to the business with respect to the cost of implementing the requirement. It turned out that a secondary requirement—the paperwork that should be eliminated had to be non-trivial—was the key economic driver. Prioritization and avoiding late changes is one reason why return on investment data is particularly useful in driving the requirements set.

Sometimes the customer is simply flighty or fickle. They change their minds because that is the way they operate. Or they simply don't know what a requirement is (you need to educate them). Other customers are simply, well, stupid. As their requirements engineer you have to tolerate a certain amount of this changeability (how much is up to you).

Finally, customers will deliberately withhold information for a variety of reasons (e.g., the information is proprietary, they distrust you, they don't like you, they don't think you will understand). The information withheld can rear its ugly head later in the project and require costly changes to the system.

Stakeholder Prioritization

Up until now, we have mostly been referring to the customer as the primary stakeholder, but of course, there are others. Not all stakeholders are of equal importance. For example, the concerns of the Baggage Handlers Union are important, but may not be as important as the airport authority (the customer), who is paying for the baggage handling system. On the other hand, federal regulations trump the desires of the customer; for example, the system must comply with all applicable federal standards.

Because we have many stakeholders and some of their needs and desires may conflict, we rank or prioritize the stakeholder classes to help resolve these situations. Usually rank denotes the risk of not satisfying the stakeholder (e.g., legal requirements should be #1 unless you want to go to jail). Ranking the stakeholders will lead to requirements prioritization, which is the key to reconciliation and risk mitigation—so get used to it!

Table 2.1 contains a partial list of stakeholders for the baggage handling system ranked in a simple High, Medium, Low priority scheme. A rationale for the rank assignment is included.

Table 2.1 Partial Ranking of Stakeholders for the Baggage Handling System

Stakeholder Class	*Rank*	*Rationale*
System maintenance personnel	Medium	They have moderate interaction with the system.
Baggage handlers	Medium	They have regular interaction with the system but have an agenda that may run counter to the customer.
Airline schedulers/ dispatchers	Low	They have little interaction with the system.
Airport personnel	Low	Most other airport personnel have little interaction with the system.
Airport managers and policy makers ("the customer")	High	They are paying for the system.

Table 2.2 Partial Ranking of Stakeholders for the Pet Store POS System

Stakeholder Class	*Rank*	*Rationale*
Cashiers	2	They have the most interaction with the system.
Managers	1	They are the primary customer/sponsor.
System maintenance personnel	4	They have to fix things when they break.
Store customers	3	They are the most likely to be adversely affected.
Inventory/warehouse personnel	6	They have the least direct interaction with the system.
Accountants/sales personnel	5	They read the reports.

Table 2.2 contains a partial list of stakeholders for the pet store POS system ranked using ratings 1 through 7, where 1 represents highest importance or priority.

You can certainly argue with this ranking and prioritizations; for example, you may think that the store customer is the most important person in the POS system. But this disagreement highlights an important point—it is early in the requirements engineering process when you want to argue about stakeholder conflicts and prioritization, not later when design decisions may have already been made that need to be undone.

Communicating with Customers and Other Stakeholders

One of the most important activities of the requirements engineer is to communicate with customers, and at times with other stakeholders. In many cases, aside from the sales personnel, the requirements engineer is the customer-facing side of the business. Therefore, it is essential that all communications be conducted clearly, ethically, consistently, and in a timely fashion.

The question arises, what is the best format for communications with customers? There are many ways to communicate, and each has specific advantages and disadvantages. For example, in-person meetings are very effective. Verbal information is conveyed via the language used, but also more subtle clues from voice quality, tone, and inflection and from body language can be conveyed. In fact, agile software methodologies advocate having a customer representative on site at all times to facilitate continuous in-person communications. Agile methodologies for requirements engineering are discussed in Chapter 7. But in-person meetings are not economical and they consume a great deal of time. Furthermore, when you have multiple customers, even geographically distributed customers, how do you meet with them? Together? Separately? Via teleconference? All of these questions have to be considered.

Well planned, intensive group meetings can be an effective form of communication for requirements engineering, and we will be discussing such techniques in the next chapter. But these meetings are expensive and time consuming and can disrupt the client's business.

Providing periodic status reports to customers during the elicitation and beyond can help to avoid some of these problems. At least from a legal standpoint, the requirements engineer has been making full disclosure of what he knows and what he does not.

Should written communications with the customer take the form of legal contracts and memoranda? The advantage here is that this kind of communication might avoid disputes, or at least provide evidence in the event of a dispute. After all, any communications with the customer can be relevant in court. But formal communications are impersonal, can slow the process of requirements engineering significantly, and can be costly (if a lawyer is involved).

Telephone or teleconference calls can be used to communicate throughout the requirements engineering process. The informality and speed of this mode is highly desirable. But even with teleconferencing, some of the nuance of co-located communication is lost, and there are always problems of misunderstanding, dropped calls, and interruptions. And the informality of the telephone call is also a liability—every communication with a customer has potential legal implications and you do not want to record every call.

Email can be effective as a means of communication, and its advantages and disadvantages fall somewhere between written memoranda and telephone calls. Email is both spontaneous and informal, but it is persistent—you can save every email transaction. But as with telephone calls, some of the interpersonal nuanced communication is lost, and as a legal document, email trails leave much to be desired.

Finally, Wiki technology can and has been used to communicate requirements information with customers and other stakeholders. The Wiki can be used as a kind of whiteboard on which ideas can be shared and refined. Further, with some massaging, the Wiki can be evolved into the final software requirements specification document. And there are ways to embed executable test cases into the SRS itself using the FitNesse acceptance testing framework (fitnesse.org).

Managing Expectations

A big part of communicating with customers is managing expectations. Expectations really matter—in all endeavors, not just requirements engineering. If you don't believe this fact, consider the following situations.

> Situation A: Imagine you were contracted to do some work as a consultant and you agreed to a fee of $5,000 for the work. You complete the work and the customer is satisfied. But your client pays you $8,000—he has had a successful year and he wants to share the wealth. How do you feel?
>
> Situation B: Now reset the clock—imagine the previous situation didn't happen yet. Imagine now that you agree to do the same work as before, but this time for $10,000. You do exactly the same amount of work as you did in situation A and the customer is satisfied. But now the customer indicates that he had a very bad year and that all he can pay is $8,000, take it or leave it. How do you feel?

In both situation A and situation B you did exactly the same amount of work and you were paid exactly the same amount of money. But you would be ecstatic in situation A, and upset in situation B. Why? What is different?

The difference is in your expectations. In situation A you expected to be paid $5,000 but the customer surprised you and exceeded your expectations, making you happy. In situation B your expectations of receiving $10,000 were not met, making you unhappy.

Some might argue that, in any endeavor, this example illustrates that you should set customers' expectations low deliberately and then exceed them, so that you can make the customer extremely happy. But this will not always work, and certainly will not work in the long run—people who get a reputation for padding schedules or otherwise low-balling expectations lose the trust of their customers and clients.*

Also recognize that you exert tremendous conscious and unconscious influence on stakeholders. When communicating with customers avoid saying such words as "I would have the system do this..." or "I don't like the system to do that ..." These phrases may influence the customer to make a decision that will be regretted later—and blamed on you!

Therefore, our goal as requirements engineers is to carefully manage customers' expectations. That is understand, adjust, monitor, reset, and then meet customer expectations at all times.

Stakeholder Negotiations

It is inevitable that along the way the requirements engineer must negotiate with customers and other stakeholders. Often the negotiations deal with convincing the customer that some desired functionality is impossible or in dealing with the

* The author often jokes when giving seminars that: "I like to set expectations deliberately low ... and then meet them."

concerns of other stakeholders. And expectation setting and management throughout the life cycle of any system project is an exercise in negotiation. While we are not about to embed a crash course in negotiation theory in this book, we wanted to mention a few simple principles that should be remembered.

Set the ground rules up front. When a negotiation is imminent, make sure that the scope and duration of the discussions are agreed. If there are to be third parties present, make sure that this is understood. If certain rules need to be followed, make both parties aware. Trying to eliminate unwanted surprises for both sides in the negotiation will lead to success.

Understand people's expectations. Make sure you realize that what matters to you might not matter to the other party. Some people care about money; others care more about their image, reputation, or feelings. When dealing with negotiations surrounding system functionality, understand what is most important to the customer. Ranking requirements will be most helpful in this regard.

Look for early successes. It always helps to build positive momentum if agreement, even on something small, can be reached. Fighting early about the most contentious issues will amplify any bad feelings and make agreement on those small issues more difficult later.

Be sure to give a little and push back a little. If you give a little in the negotiation, it always demonstrates good faith. But the value of pushing back in a negotiation is somewhat counterintuitive. It turns out that by not pushing back, you leave the other party feeling cheated and empty.

For example, suppose someone advertises a used car for sale at $10,000. You visit the seller, look at the car, and offer $8,000. The seller immediately accepts. How do you feel? You probably feel that the seller accepted too easily, and that he had something to hide. Or, you feel that the $10,000 asking price was grossly inflated—why, if you had offered $10,000, the seller would have accepted that—how greedy of him? You would have actually felt better if the seller refused your offer of $8,000 but countered with $9,000 instead. So, push back a little.

Conclude negotiating only when all parties are satisfied. Never end a negotiation with open questions or bad feelings. Everyone needs to feel satisfied and whole at the end of a negotiation. If you do not ensure mutual satisfaction, you will likely not do business together again, and your reputation at large may be damaged (customers talk to one another).

There are many good texts on effective negotiation (e.g., Cohen 2000), and it is advisable that all requirements engineers continuously practice and improve their negotiating skills.

Exercises

2.1. Why is it important to have a mission statement?

2.2. What is the relationship of a system mission statement to the enterprise's mission statement?

2.3. When should a domain vocabulary be established?

2.4. Under what circumstances might the customer's needs and desires be considered secondary?
2.5. Think of a system or product you use often and try to guess what the mission statement is for that product, or the company that makes it.
2.6. Search the Internet for the actual mission statement and compare it to your guess. Does such a mission statement exist? If it does exist, how does the actual mission statement compare to your guess?
2.7. At what stage of the requirements development are additions to the requirements considered scope creep?

References

Cohen, H. (2000) *You Can Negotiate Anything*, Citadel.
Maslow, A. (1943) A theory of human motivation, *Psychological Review,* 50: 370–396.
Valacich, J.H., D.V. Parboteeah, and J.D. Wells (2007) The Online Consumer's Hierarchy of Needs, *Communications of the ACM*, 50(9): 84–90.

Chapter 3

Requirements Elicitation

Introduction

In this chapter we explore the many ways that requirements can be elicited—that is, found, discovered, captured, coerced, whatever the term may be. Remember that requirements are usually not so easy to come by, at least not all of them. Many of the more subtle and complex ones have to be teased out through rigorous, if not, dogged processes.

There are many techniques that you can choose to conduct requirements elicitation, and you will probably need to use more than one, and perhaps different ones for different classes of users/stakeholders. The techniques that we will discuss are

- Brainstorming
- Card Sorting
- Designer as Apprentice
- Domain Analysis
- Ethnographic Observation
- Goal-Based Approaches
- Group Work
- Interviews
- Introspection
- Joint Application Development (JAD)
- Laddering
- Protocol Analysis
- Prototyping

- Quality Function Deployment (QFD)
- Questionnaires
- Repertory Grids
- Scenarios
- Task Analysis
- User Stories
- Viewpoints
- Workshops

This list is based on one offered by Zowghi and Coulin (1998).

Elicitation Techniques Survey

Now it is time to begin examining the elicitation techniques. We offer these techniques in alphabetical order—no preference is implied. At the end of the chapter, we will discuss the prevalence and suitability of these techniques in different situations.

Brainstorming

Brainstorming consists of informal sessions with customers and other stakeholders to generate overarching goals for the systems. Brainstorming can be formalized to include a set agenda, minutes taking, and the use of formal structures (e.g., Roberts Rules of Order). But the formality of a brainstorming meeting is probably inversely proportional to the creative level exhibited at the meeting. These kinds of meetings probably should be informal, even spontaneous, with the only structure embodying some recording of any major discoveries.

During brainstorming sessions, some preliminary requirements may be generated, but this aspect is secondary to the process. The JAD technique incorporates brainstorming (and a whole lot more), and it is likely that most other group-oriented elicitation techniques embody some form of brainstorming implicitly. Brainstorming is also useful for general objective setting, such as mission or vision statement generation.

Card Sorting

This technique involves having stakeholders complete a set of cards that includes key information about functionality for the system/software product. It is also a good idea for the stakeholders/customers to include ranking and rationale for each of the functionalities.

The period of time to allow customers and stakeholders to complete the cards is an important decision. While the exercise of card sorting can be completed in a few

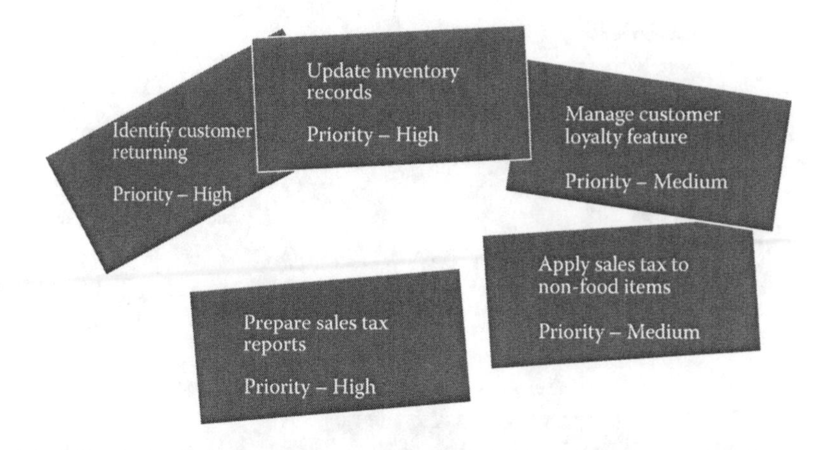

Figure 3.1 A tiny subset of the unsorted cards generated by customers for the pet store POS system.

hours, rushing the stakeholders will likely lead to important, missing functionalities. Giving stakeholders too much time, however, can slow the process unnecessarily. It is recommended that a minimum of one week (and no more than two weeks) be allowed for the completion of the cards. Another alternative is to have the customers complete the cards in a two-hour session, then return one week later for another session of card completion and review.

In any case, after each session of card generation the requirements engineer organizes these cards in some manner, generally clustering the functionalities logically. These clusters form the bases of the requirements set. The sorted cards can also be used as an input to the process to develop CRC (capability, responsibility, class) cards to determine program classes in the eventual code. Another technique to be discussed shortly, QFD, includes a card-sorting activity.

To illustrate the process, Figure 3.1 depicts a tiny subset of cards generated by the customer for the pet store POS system, lying in an unsorted pile. In this case, each card contains only a brief description of the functionality and a priority rating is included (no rationale is shown for brevity).

The requirements engineer analyzes this pile of cards and decides that two of the cards pertain to "customer management" functions, two cards to "tax functions," and one card to "inventory features" or functions, and arranges the cards in appropriate piles as shown in Figure 3.2.

The customer can be shown this sorted list of functionalities for correction or missing features. Then, a new round of cards can be generated if necessary. The process continues until the requirements engineer and customer are satisfied that the system features are substantially captured.

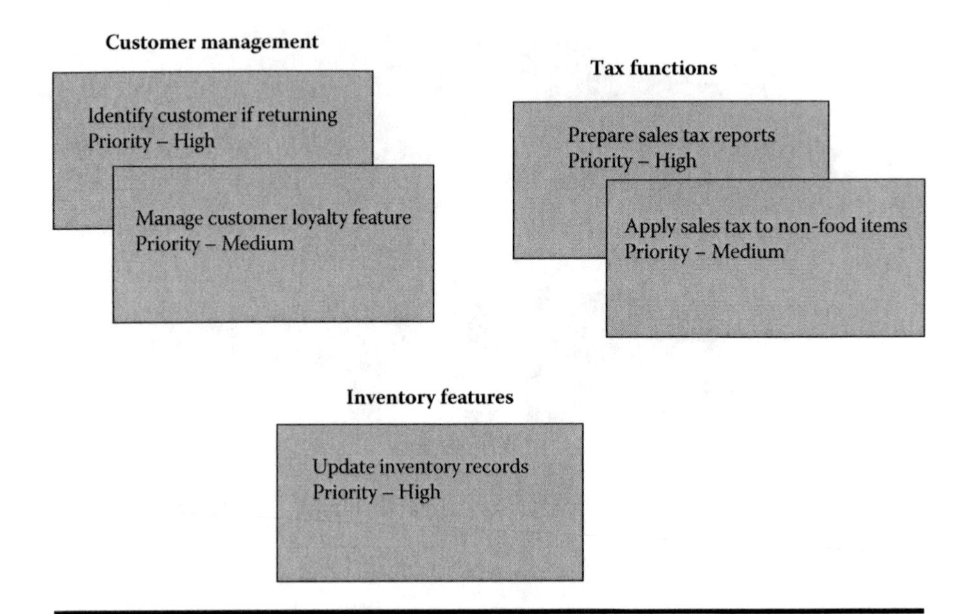

Figure 3.2 Sorted cards for the pet store POS system.

*Designer as Apprentice**

Designer as apprentice is a requirements discovery technique in which the requirements engineer "looks over the shoulder" of the customer in order to learn enough about the customer's work to understand their needs. The relationship between customer and designer is like that between a master craftsman and apprentice. That is, the apprentice learns a skill from the master just as we want the requirements engineer (the designer) to learn about the customer's work from the customer. The apprentice is there to learn whatever the master knows (and therefore must guide the customer in talking about and demonstrating those parts of the work). The designer is there to address specific needs.

It might seem that the customer needs to have some kind of teaching ability for this technique to work, but that is not true. Some customers cannot talk about their work effectively, but can talk about it as it unfolds. Moreover, customers don't have to work out the best way to present it, or the motives; they just explain what they're doing.

Seeing the work reveals what matters. For example, people are not aware of everything they do and sometimes why they do it. Some actions are the result of years of experience and are too subtle to express. Other actions are just habits with no valid justification. The presence of an apprentice provides the opportunity for the master (customer) to think about the activities and how they came about.

* This discussion is adapted from one found in Laplante (2006), with permission.

Seeing the work reveals details since, unless we are performing a task, it is difficult to be detailed in describing it. Finally, seeing the work reveals structure. Patterns of working are not always obvious to the worker. An apprentice learns the strategies and techniques of work by observing multiple instances of a task and forming an understanding of how to do it themselves, incorporating the variations.

In order for this technique to work, the requirements engineer must understand the structure and implication of the work, including

- the strategy to get work done
- constraints that get in the way
- the structure of the physical environment as it supports work
- the way work is divided
- recurring patterns of activity
- the implications these have on any potential system

The designer must demonstrate an understanding of the work to the customer so that any misunderstandings can be corrected.

Finally, using the designer as apprentice approach provides other project benefits beyond requirements discovery. For example, using this technique can help improve the process that is being modeled.

Both customer and designer learn during this process—the customer learns what may be possible and the designer expands his understanding of the work. If the designer has an idea for improving the process, however, this must be fed back to the customer immediately (at the time).

Domain Analysis

We have already emphasized the importance of having domain knowledge (whether it is had by the requirements engineer and/or the customer) in requirements engineering. Domain analysis involves any general approach to assessing the "landscape" of related and competing applications to the system being designed. Such an approach can be useful in identifying essential functionality and, later, missing functionality. Domain analysis can also be used downstream for identifying reusable components (such as open source software elements that can be incorporated into the final design). The QFD elicitation approach explicitly incorporates domain analysis, and we will discuss this technique shortly.

Ethnographic Observation

Ethnographic observation refers to any technique in which observation of indirect and direct factors inform the work of the requirements engineer. Ethnographic observation is a technique borrowed from social science in which observations of human activity and the environment in which the work occurs are used to inform the scientist in the study of some phenomenon. In the strictest sense, ethnographic observation involves long periods of observation (hence, an objection to its use as a requirements elicitation technique).

To illustrate ethnographic observation, imagine the societal immersion of an anthropologist studying some different culture. The anthropologist lives among the culture being studied, but in a way in which he is minimally intrusive. While eating, sleeping, hunting, celebrating, mourning, and so on within the culture, all kinds of direct and indirect evidence of how that society functions and its belief systems are collected.

In applying ethnographic observation to requirements elicitation, the requirements engineer immerses himself in the workplace culture of the customer. Here, in addition to observing work or activity to be automated, the requirements engineer is also in a position to collect evidence of customer needs derived from the surroundings that may not be communicated directly. Designer as apprentice is one requirements elicitation technique that includes the activity of ethnographic observation.

To illustrate this technique in practice, consider this situation in which ethnographic observation occurs:

- You are gathering requirements for a Smart Home for some customer.
- You spend long periods of time interviewing the customer about what she wants.
- You spend time interacting with the customer as she goes about her day and ask questions ("why are you running the dishwasher at night, why not in the morning?").
- You spend long periods of time passively observing the customer "in action" in her current home to get nonverbal clues about her wants and desires.
- You gain other information from the home itself—the books on the bookshelf, paintings on the wall, furniture styles, evidence of hobbies, signs of wear and tear on various appliances, etc.

Ethnographic observation can be very time-consuming and requires substantial training of the observer to be useful. There is another objection, based on the intrusiveness of the process. There is a well-known principle in physics known as the Heisenberg uncertainty principle, which, in layperson's terms, means that you can't precisely measure something without affecting that which you are measuring. So, for example, when you are observing the work environment for a client, processes and behaviors change because everyone is out to impress—so an incorrect picture of the situation is formed, leading to flawed decisions down the line.

Goal-Based Approaches

Goal-based approaches comprise any elicitation techniques in which requirements are recognized to emanate from the mission statement, through a set of goals that lead to requirements. That is, looking at the mission statement, a set of goals that fulfill that mission is generated. These goals may be subdivided one or more times to obtain lower-level goals. Then, the lower-level goals are branched out into specific high-level requirements. Finally, the high-level requirements are used to generate lower-level ones.

For example, consider the baggage handling system mission statement:

> To automate all aspects of baggage handling from passenger origin to destination.

The following goals might be considered to fulfill this mission:

- Goal 1: To completely automate the tracking of baggage from check-in to pick-up.
- Goal 2: To completely automate the routing of baggage from check-in counter to plane.
- Goal 3: To reduce the amount of lost luggage to 1%.

Related to this process of goal decomposition into requirements is a technique called "Goal-Question-Metric" (GQM). GQM is used in the creation of metrics that can be used to test requirements satisfaction. This process primarily comprises three steps: state the goals that the organization is trying to achieve; next, derive from each goal the questions that must be answered to determine if they are being met; finally, decide what must be measured in order to be able to answer the questions (Basili and Weiss 1984).

So, in the case of the baggage handling system, consider goal 3. Here the related question is "What percentage of luggage is lost for a given [airport/airline/flight/time period/etc.]?" This question suggests a requirement of the form

> The percentage of luggage lost for a given [airport/airline/flight/time period/etc.] shall be not greater than 1%.

The associated metric for this requirement, then, is simply the percentage of luggage lost for a particular [airport/airline/flight/time period/etc.]. Of course, we really need a definition for "lost luggage," since so-called lost luggage often reappears days or even months after it is declared lost. Also, reasonable assumptions need to be made in framing this requirement in terms of an airport's reported luggage losses over some time period, or for a particular airline at some terminal, and so forth.

In any case, we deliberately picked a simple example here—the appropriate question for some goal (requirement) is not always so obvious, nor is the associated metric so easily derived from the question.

Group Work

Group work is a general term for any kind of group meetings that are used during the requirements discovery, analysis, and follow-up processes. The most celebrated of group-oriented work for requirements elicitation is Joint Application Design (JAD), which we will discuss shortly.

Group activities can be very productive in terms of bringing together many stakeholders, but risk the potential for conflict and divisiveness. The key to success in any kind of group work is in the planning and execution of the group meetings. Here are the most important things to remember about group meetings.

- Do your homework—research all aspects of the organization, problems, politics, environment, and so on.
- Publish an agenda (with time allotted for each item) several days before the meeting occurs.
- Stay on the agenda throughout the meeting (no meeting scope creep).
- Have a dedicated note-taker (scribe) on hand.
- Do not allow personal issues to creep in.
- Allow all to have their voices heard.
- Look for consensus at the earliest opportunity.
- Do not leave until all items on the agenda have received sufficient discussion.
- Publish the minutes of the meeting within a couple of days of meeting close and allow attendees to suggest changes.

These principles will come into play for the JAD approach to requirements elicitation.

Group work of any kind has many drawbacks. First, group meetings can be difficult to organize and get the many stakeholders involved to focus on issues. Problems of openness and candor can occur as well because people are not always eager to express their true feelings in a public forum. Because everyone has a different personality, certain individuals can dominate the meeting (and these may not be the most "important" individuals). Allowing a few to own the meeting can lead to feelings of being "left out" for many of the other attendees.

Running effective meetings, and hence using group work, requires highly developed leadership, organizational, and interpersonal skills. Therefore, the requirements engineer should seek to develop these skills whenever possible.

Interviews

The "opposite" of group activities is the one-on-one (or small group) interview. This is an obvious and easy-to-use technique to extract system requirements from a customer.

There are three kinds of interviews that can be used in elicitation activities:

- unstructured
- structured
- semi-structured

Unstructured interviews, which are probably the most common type, are conversational in nature and serve to relax the customer. Like a spontaneous "confession" these can occur any time and any place whenever the requirements engineer and customer are together, and the opportunity to capture information this way should never be lost. But depending on the skill of the interviewer, unstructured interviews can be hit or miss. Therefore, structured or semi-structured interviews are preferred.

Structured interviews are much more formal in nature, and they use predefined questions that have been rigorously planned. Templates are very helpful when employed with interviewing using the structured style. The main drawback to structured interviews is that some customers may withhold information because the format is too "stodgy."

Semi-structured interviews combine the best of structured and unstructured interviews. That is, the requirements engineer prepares a carefully thought-out list of questions, but then allows for spontaneous unstructured questions to creep in during the course of the interview.

While structured interviews are preferred, the choice of which one to use is very much an opportunistic decision. For example, when the client's corporate culture is very informal and relaxed, and trust is high, then unstructured interviews might be preferred. In a stodgier, process-oriented organization, structured and semi-structured interviews are probably more desirable.

Here are some sample interview questions that can be used in any of the three interview types.

- Name an essential feature of the system? Why is this feature important?
- On a scale of one to five, five being most important, how would you rate this feature?
- How important is this feature with respect to other features?
- What other features are dependent on this feature?
- What other features must be independent of this feature?
- What other observations can you make about this feature?

Whatever interview technique is used, care must be taken to ensure that all of the right questions are asked. That is, leave out no important questions, and include no extraneous, offensive, or redundant questions. When absolutely necessary, interviews can be done via telephone, videoconference, or email, but be aware that, in these modes of communication, certain important nuanced aspects to the responses may be lost.

Introspection

When a requirements engineer develops requirements based on what he "thinks" the customer wants, then he is conducting the process of introspection. In essence the requirements engineer puts himself in the place of the customer and opines "if I were the customer I would want the system to do this ..."

An introspective approach is useful when the requirements engineer's domain knowledge far exceeds the customer's. Occasionally, the customer will ask the engineer questions similar to the following—"if you were me, what would you want?" While introspection will inform every aspect of the requirements engineer's interactions, remember our admonition about not telling a customer what he ought to want.

Joint Application Design (JAD)*

Joint Application Design (JAD) involves highly structured group meetings or mini-retreats with system users, system owners, and analysts in a single venue for an extended period of time. These meetings occur four to eight hours per day and over a period lasting one day to a couple of weeks.

JAD and JAD-like techniques are becoming increasingly common in systems planning and systems analysis to obtain group consensus on problems, objectives, and requirements. Specifically, software engineers can use JAD for

- eliciting requirements and for the software requirements specification
- design and software design description
- code
- tests and test plans
- users' manuals

There can be multiple reviews for each of these artifacts, if necessary. But JAD reviews are especially important as a requirements elicitation tool.

Planning for a JAD review or audit session involves three steps:

1. selecting participants
2. preparing the agenda
3. selecting a location

Great care must be taken in preparing each of these steps.

Reviews and audits may include some or all of the following participants:

- sponsors (for example, senior management)
- a team leader (facilitator, independent)

---

* Ibid.

- users and managers who have ownership of requirements and business rules
- scribes
- engineering staff

The sponsor, analysts, and managers select a leader. The leader may be in-house or a consultant. One or more scribes (note takers) are selected, normally from the software development team. The analyst and managers must select individuals from the user community. These individuals should be knowledgeable and articulate in their business area.

Before planning a session, the analyst and sponsor must determine the scope of the project and set the high-level requirements and expectations of each session. The session leader must also ensure that the sponsor is willing to commit people, time, and other resources to the effort. The agenda depends greatly on the type of review to be conducted and should be constructed to allow for sufficient time. The agenda, code, and documentation must also be sent to all participants well in advance of the meeting so that they have sufficient time to review them, make comments, and prepare to ask questions.

The following are some rules for conducting software requirements, design audits, or code walkthrough. The session leader must make every effort to ensure these practices are implemented.

- Stick to agenda.
- Stay on schedule (agenda topics are allotted specific time).
- Ensure that the scribe is able to take notes.
- Avoid technical jargon (if the review involves nontechnical personnel).
- Resolve conflicts (try not to defer them).
- Encourage group consensus.
- Encourage user and management participation without allowing individuals to dominate the session.
- Keep the meeting impersonal.
- Allow the meetings to take as long as necessary.

The end product of any review session is typically a formal written document providing a summary of the items (specifications, design changes, code changes, and action items) agreed upon during the session. The content and organization of the document obviously depend on the nature and objectives of the session. In the case of requirements elicitation, however, the main artifact could be a first draft of the SRS.

Laddering

In laddering, the requirements engineer asks the customer short prompting questions ("probes") to elicit requirements. Follow-up questions are then posed to dig

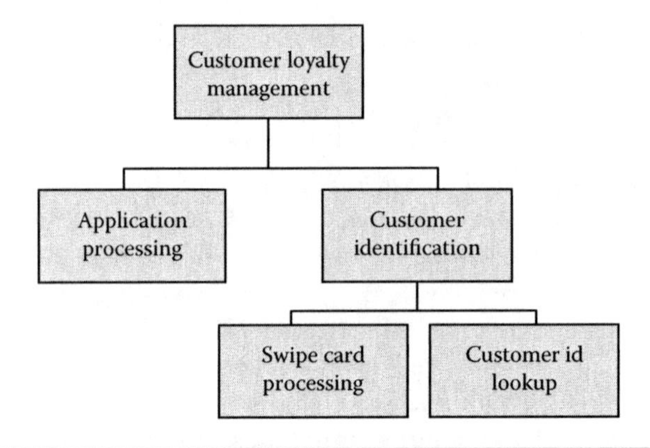

Figure 3.3 Laddering diagram for the pet store POS system.

deeper below the surface. The resultant information from the responses is then organized into a tree-like structure.

To illustrate the technique, consider the following sequence of laddering questions and responses for the pet store POS system. "RE" refers to the requirements engineer.

RE: Name a key feature of the system?
Customer: Customer identification.
RE: How do you identify a customer?
Customer: They can swipe their loyalty card.
RE: What if a customer forgets their card?
Customer: They can be looked up by phone number.
RE: When do you get the customer's phone number?
Customer: When customers complete the application for the loyalty card.
RE: How do customers complete the applications? …

And so on. Figure 3.3 shows how the responses to the questions are then organized in a ladder or hierarchical diagram.

The laddering technique assumes that information can be arranged in a hierarchical fashion, or, at least, it causes the information to be arranged hierarchically.

Protocol Analysis

A protocol analysis is a process where customers, together with the requirements engineers, walk through the procedures that they are going to automate. During such a walk-through, the customers explicitly state the rationale for each step that is being taken.

While you will see shortly that this technique is very similar to designer as apprentice, there are subtle differences. These differences lie in the role of the requirements engineer who is more passive in protocol analysis than in designer as apprentice.

Prototyping

Prototyping involves construction of models of the system in order to discover new features. Prototypes can involve working models and nonworking models. Working models can include working code in the case of software systems and simulations or temporary or to-scale prototypes for non-software systems. Nonworking models can include storyboards and mock-ups of user interfaces. Building architects use prototypes regularly (e.g., scale drawings, cardboard models, 3-D computer animations) to help uncover and confirm customer requirements. Systems engineers use prototypes for the same reasons.

In the case of working code prototypes, the code can be deliberately designed to be throwaway or it can be deliberately designed to be reused (non-throwaway). For example, graphical user interface code mock-ups can be useful for requirements elicitation and the code can be reused. And most agile methodologies incorporate a process of continuously evolving non-throwaway prototypes.

In unfortunate cases, prototypes that were not intended to be kept, are in fact kept because of schedule pressures. This situation is rather unfortunate, since the code was likely not designed using the most rigorous techniques, but is all too commonly found in industry.

Prototyping is a particularly important technique for requirements elicitation. It is used extensively, for example, in the spiral software development model, and agile methodologies consist essentially of a series of increasingly functional non-throwaway prototypes.

To see how widely prototyping is used, consider the summary of the responses to a question concerning the use of prototyping in the 2003 survey previously mentioned, which is shown in Figure 3.4 (Neill and Laplante 2003).

Here we see that 60% of respondents said that their organizations used prototyping for requirements elicitation. It is noteworthy that more than two-thirds of respondents did do prototyping and that 4% did not know if their organization did prototyping.

Finally, the survey authors sought to determine what kind of prototyping companies were using. There are a number of different ways to use prototyping—for example, within a fourth-generation environment (that is, a simulator), throw-away prototyping, evolutionary prototyping (where the prototype evolves into the final

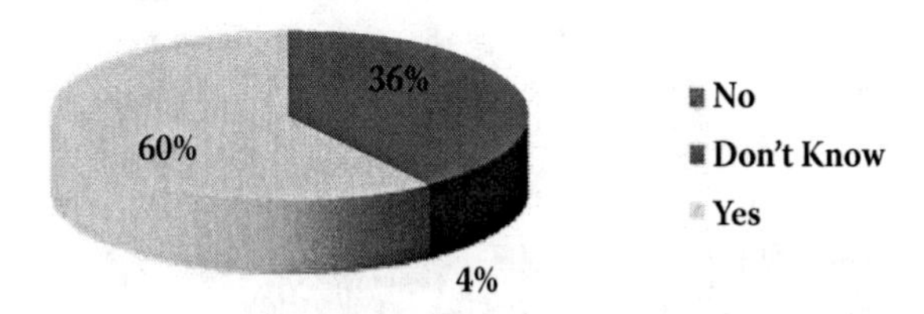

Figure 3.4 Does your company do prototyping? Neill-Laplante survey of Southeast PA software professionals, 186 respondents (Neill and Laplante 2003).

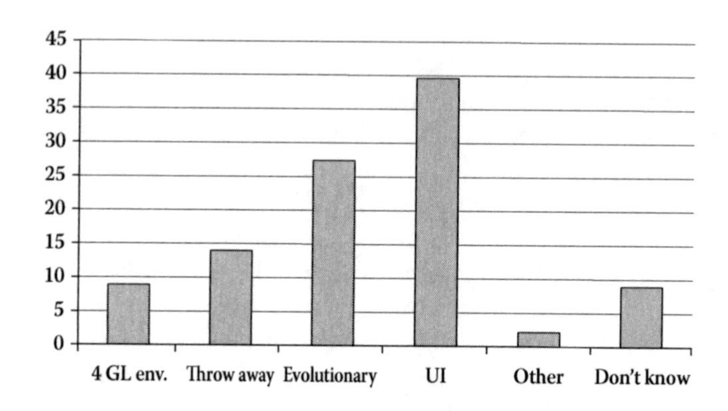

Figure 3.5 What type of prototyping was performed? Neill-Laplante survey, 186 responses (Neill and Laplante 2003).

system), or user interface (UI) prototyping—and the survey results show the breakdown of these responses (Figure 3.5).

Here it was surprising that 27% of those who used prototyping did evolutionary prototyping. Evolutionary prototyping is considered dangerous because prototypes are not ordinarily designed for release, yet the code finds its way into the release. However, as previously noted, agile methodologies embody evolutionary prototyping, and so some of these respondents may have been referring to the use of prototypes in that setting.

*Quality Function Deployment**

Quality function deployment (QFD) is a technique for discovering customer requirements and defining major quality assurance points to be used throughout the production phase. QFD provides a structure for ensuring that customers' needs and desires are carefully heard, then directly translated into a company's internal technical requirements—from analysis through implementation to deployment. The basic idea of QFD is to construct relationship matrices between customer needs, technical requirements, priorities, and (if needed) competitor assessment. In essence, QFD incorporates card sorting and laddering and domain analysis.

Because these relationship matrices are often represented as the roof, ceiling, and sides of a house, QFD is sometimes referred to as the "house of quality" (Figure 3.6; Akao 1990).

QFD was introduced by Yoji Akao in 1966 for use in manufacturing, heavy industry, and systems engineering. It has also been applied to software systems by IBM, DEC, HP, AT&T, Texas Instruments, and others.

---

* Ibid.

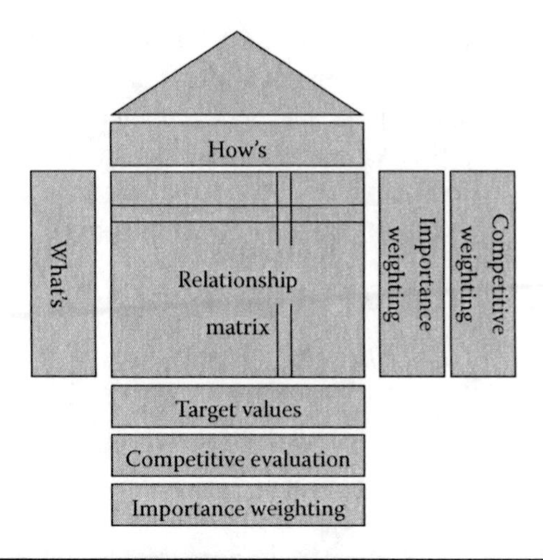

Figure 3.6 QFD's "house of quality" (Akao 1990).

When we refer to the "voice of the customer" we mean that the requirements engineer must empathically listen to customers to understand what they need from the product, as expressed by the customer in their words. The voice of the customer forms the basis for all analysis, design, and development activities, to ensure that products are not developed from only "the voice of the engineer." This approach embodies the essence of requirements elicitation.

QFD has several advantages. QFD improves the involvement of users and managers. It shortens the development lifecycle and improves overall project development. QFD supports team involvement by structuring communication processes. Finally, it provides a preventive tool that avoids the loss of information.

There are some drawbacks to QFD, however. For example, there may be difficulties in expressing temporal requirements. And QFD is difficult to use with an entirely new project type. For example, how do you discover customer requirements for something that does not exist and how do you build and analyze the competitive products? In these cases the solution is to look at similar or related products, but still there is apt to be a cognitive gap.

Finally, sometimes it is hard to find measurements for certain functions and to keep the level of abstraction uniform. And, the less we know the less we document. Finally, as the feature list grows uncontrollably, the house of quality can become a "mansion."

Questionnaires

Requirements engineers often use questionnaires and other survey instruments to reach large groups of stakeholders. Surveys are generally used at early stages of the elicitation process to quickly define the scope boundaries.

Survey questions of any type can be used. For example, questions can be closed- (e.g., multiple choice, true-false) or open-ended—involving free-form responses. Closed questions have the advantage of easier coding for analysis, and they help to bound the scope of the system. Open questions allow for more freedom and innovation, but can be harder to analyze and can encourage scope creep.

For example, some possible survey questions for the pet store POS system are

- How many unique products (SKUs) do you carry in your inventory?

  a) 0-1000 b) 1001-10,000 c) 10,001-100,000 d) >100,000

- How many different warehouse sites do you have? ____
- How many different store locations do you have? ____
- How many unique customers do you currently have? ____

There is a danger in over-scoping and under-scoping if questions are not adequately framed, even for closed-ended questions. Therefore, survey elicitation techniques are most useful when the domain is very well understood by both stakeholders and requirements engineer.

Surveys can be conducted via telephone, email, in person, and using Web-based technologies. There are a variety of commercial tools and open source solutions that are available to simplify the process of building surveys and collecting and analyzing results that should be employed.

Repertory Grids

Repertory grids incorporate a structured ranking system for various features of the different entities in the system and are typically used when the customers are domain experts. Repertory grids are particularly useful for identification of agreement and disagreement within stakeholder groups.

The grids look like a feature or quality matrix in which rows represent system entities and desirable qualities and columns represent rankings based on each of the stakeholders. While the grids can incorporate both qualities and features, it is usually the case that the grids have all features or all qualities to provide for consistency of analysis and dispute resolution.

To illustrate the technique, Figure 3.7 represents a repertory grid for various qualities of the baggage handling system. Here we see that for the airport operations manager, all qualities are essentially of highest importance (safety is rated as slightly lower, at 4). But for the Airline Worker's Union representative, safety is the most important (after all, his union membership has to interact with the system on a daily basis).

In essence, these ratings reflect the agendas or differing viewpoints of the stakeholders. Therefore, it is easy to see why the use of repertory grids can be very helpful in confronting disputes involving stakeholder objectives early. In addition, the grids

	Airline worker's union rep	Maintenance engineer	Airport operations manager
Baggage handling speed	1	1	5
Fault-tolerance	4	5	5
Safety	5	4	4
Reliability	3	5	5
Ease of maintenance	3	5	5

1=lowest importance

Figure 3.7 Partial repertory grid for the baggage handling system.

can provide valuable documentation for dealing with disagreements later in the development of the system because they capture the attitudes of the stakeholders about qualities and features in a way that is hard to dismiss.

Scenarios

Scenarios are informal descriptions of the system in use that provide a high-level description of system operation, classes of users, and exceptional situations.

Here is a sample scenario for the pet store POS system.

> A customer walks into the pet store and fills his cart with a variety of items. When he checks out, the cashier asks if the customer has a loyalty card. If he does, she swipes the card, authenticating the customer. If he does not, then she offers to complete one for him on the spot.
>
> After the loyalty card activity, the cashier scans products using a bar code reader. As each item is scanned, the sale is appropriately totaled and the inventory is appropriately updated. Upon completion of product scanning a subtotal is computed. Then any coupons and discounts are entered. A new subtotal is computed and applicable taxes are added. A receipt is printed and the customer pays using cash, credit card, debit card, or check. All appropriate totals (sales, tax, discounts, rebates, etc.) are computed and recorded.

Scenarios are quite useful when the domain is novel (consider a scenario for the Space Station, for example). User stories are, in fact, a form of scenario.

Task Analysis

Like many of the hierarchically oriented techniques that we have studied already, task analysis involves a functional decomposition of tasks to be performed by the

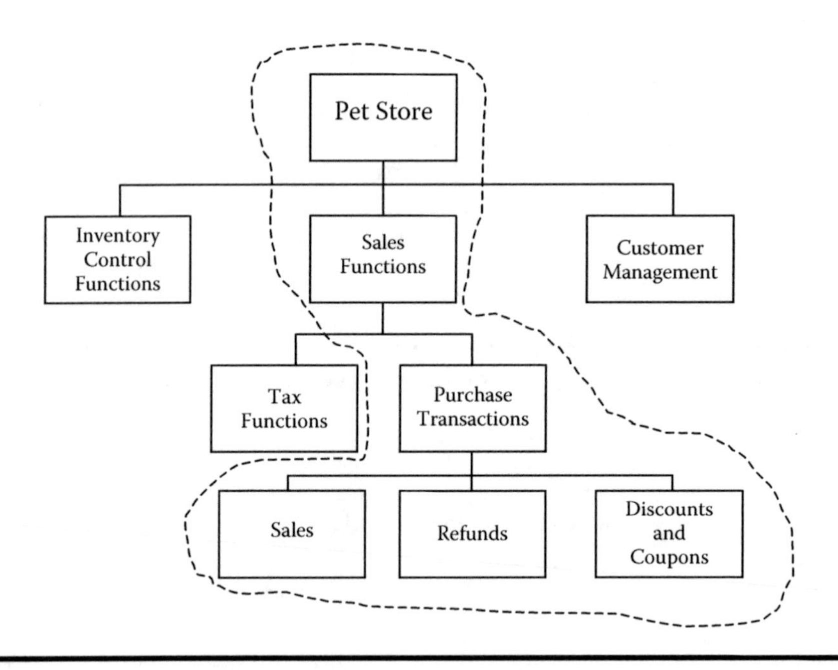

Figure 3.8 Partial task analysis for the pet store POS system.

system. That is, starting at the highest level of abstraction, the designer and customers elicit further levels of detail. This detailed decomposition continues until the lowest level of functionality (single task) is achieved.

As an example, consider the partial task analysis for the pet store POS system shown in Figure 3.8. Here the overarching "pet store POS system" is deemed to consist of three main tasks: inventory control, sales, and customer management. Drilling down under the sales functions, we see that these consist of the following tasks: tax functions and purchase transactions. Next proceeding to the purchase transaction function, we decompose these tasks into sales, refunds, and discounts and coupons tasks.

The task analysis and decomposition continues until a sufficient level of granularity is reached (typically, to the level of a method or nondecomposable procedure) and the diagram is completed.

*User Stories**

User stories are short conversational text that are used for initial requirements discovery and project planning. User stories are widely employed in conjunction with agile methodologies.

* Ibid.

User stories are written by the customers in terms of what the system needs to do for them and in their own "voice." User stories usually consist of two to four sentences written on a three-by-five-inch card. About 80 user stories is said to be appropriate for one system increment or evolution, but the appropriate number will vary widely depending on the application size and scope and development methodology to be used (e.g., agile versus incremental).

An example of a user story for the pet store POS system is as follows:

Each customer should be able to easily check out at a register.
Self-service shall be supported.
All coupons, discounts, and refunds should be handled this way.

User stories should only provide enough detail to make a reasonably low-risk estimate of how long the story will take to implement. When the time comes to implement, the story developers will meet with the customer to flesh out the details.

User stories also form the basis of acceptance testing. For example, one or more automated acceptance tests can be created to verify the user story has been correctly implemented.

Viewpoints

Viewpoints are a way to organize information from the (point of view of) different constituencies. For example, in the baggage handling system, there are differing perspectives of the system for each of the following stakeholders:

- baggage handling personnel
- travelers
- maintenance engineers
- airport managers
- regulatory agencies

By recognizing the needs of each of these stakeholders and the contradictions raised by these viewpoints, conflicts can be reconciled using various approaches.

The actual viewpoints incorporate a variety of information from the business domain, process models, functional requirements specs, organizational models, etc.

Sommerville and Sawyer (1997) suggested the following components should be in each viewpoint:

- a representation style, which defines the notation used in the specification
- a domain, which is defined as "the area of concern addressed by the viewpoint"
- a specification, which is a model of a system expressed in the defined style

- a work plan, with a process model, which defines how to build and check the specification
- a work record, which is a trace of the actions taken in building, checking and modifying the specification

Viewpoint analysis is typically used for prioritization, agreement, and ordering of requirements.

Workshops

On a most general level, workshops are any formal or informal gathering of stakeholders to hammer out requirements issues. We can distinguish workshops as being of two types, formal and informal.

Formal workshops are well-planned meetings and are often "deliverable" events that are mandated by contract. For example, DOD MIL STD 2167 incorporated multiple required and optional workshops (critical reviews). A good example of a formal workshop style is embodied in JAD.

Informal workshops are usually less boring than highly structured meetings. But informal meetings tend to be too sloppy and may lead to a sense of false security and lost information. If some form of workshop is needed it is recommended that a formal one be held using the parameters for successful meetings previously discussed.

Elicitation Summary

This tour has included many elicitation techniques, and each has its advantages and disadvantages, which were discussed along the way. Clearly, some of these techniques are too general, some too specific, some rely too much on stakeholder knowledge, some not enough, etc. Therefore, it is clear that some combination of techniques is needed to successfully address the requirements elicitation challenge.

Which Combination of Requirements Elicitation Techniques Should Be Used?

In order to facilitate the discussion about appropriate elicitation techniques, we can roughly cluster the techniques previously discussed into categories or equivalence classes (interviews, domain-oriented, group-work, ethnography, prototyping, goals, scenarios, viewpoints) as shown in Table 3.1.

Now we can summarize how effective various techniques are in dealing with various aspects of the elicitation process as shown in Table 3.2 (based on work by Zowghi and Coulin 1998).

Table 3.1 Organizing Various Elicitation Techniques Roughly by Type

Technique Type	*Techniques*
Domain-oriented	Card sorting Designer as apprentice Domain analysis Laddering Protocol analysis Task analysis
Ethnography	Ethnographic observation
Goals	Goal-based approaches QFD
Group work	Brainstorming Group work JAD Workshops
Interviews	Interviews Introspection Questionnaires
Prototyping	Prototyping
Scenarios	Scenarios User stories
Viewpoints	Viewpoints Repertory grids

Source: Zowghi and Coulin (1998).

For example, interview-based techniques are useful for all aspects of requirements elicitation (but are very time-consuming). On the other hand, prototyping techniques are best used to analyze stakeholders and to elicit the requirements. Ethnographic techniques are good for understanding the problem domain, analyzing stakeholders, and soliciting requirements. And so on.

Finally, there is clearly overlap between these elicitation techniques (clusters) in that some accomplish the same thing and, hence, are alternatives to each other. In other cases, these techniques complement one another. In Table 3.3 alternative (A) and complementary (C) elicitation groupings are shown.

Table 3.2 Techniques and Approaches for Elicitation Activities

	Interviews	*Domain*	*Groupwork*	*Ethnography*	*Prototyping*	*Goals*	*Scenarios*	*Viewpoints*
Understanding the domain	•	•	•	•		•	•	•
Identifying sources of requirements	•	•	•			•	•	•
Analyzing the stakeholders	•	•	•	•	•	•	•	•
Selecting techniques and approaches	•	•	•					
Eliciting the requirements	•	•	•	•	•	•	•	•

Source: Zowghi and Coulin (1998).

Table 3.3 Complementary and Alternative Techniques

	Interviews	*Domain*	*Groupwork*	*Ethnography*	*Prototyping*	*Goals*	*Scenarios*	*Viewpoints*
Interviews		C	A	A	A	C	C	C
Domain	C		C	A	A	A	A	A
Groupwork	A	C		A	C	C	C	C
Ethnography	A	A	A		C	C	A	A
Prototyping	A	A	C	C		C	C	C
Goals	C	A	C	C	C		C	C
Scenarios	C	A	C	A	C	C		A
Viewpoints	C	A	C	A	C	C	A	

Source: Zowghi and Coulin (1998).

Clearly, in selecting a set of techniques to be used, the requirements engineer would look for a set of complementary techniques. For example, a combination of viewpoint analysis and some form of prototyping would be desirable. On the other hand using both viewpoint analysis and scenario generation would probably yield excessively redundant information.

There is no "silver bullet" combination of elicitation techniques. The right mix will depend on the application domain, the culture of the customer organization and that of the requirements engineer, the size of the project, and many other factors. You can use Tables 3.1 through 3.3 to guide you through selecting an appropriate set of elicitation techniques.

Prevalence of Requirements Elicitation Techniques

Before we conclude this discussion, let's get an idea of how various elicitation techniques are commonly used in industry. To do so, we return to the survey of 2003 (Neill and Laplante). A summary of the answers to the question "which requirements elicitation technique(s) do you use?" is shown in Figure 3.9.

The data revealed that over 50% surveyed used scenarios or use cases in the requirements phase; contrast this with the fact that a subsequent question revealed that object-orientation, a technique often applied in conjunction with use cases, was only reported by 30% of the survey population. Other popular approaches to requirements elicitation reported included group-consensus-type techniques such as use case diagrams, JAD, interviews, and focus groups (Neill and Laplante 2003).

Elicitation Support Technologies

We close this chapter by remarking on some technologies that can be used to support various requirements elicitation processes and techniques previously discussed. These technologies include

- Wikis
- Mobile technologies
- Content analysis

Using Wikis for Requirements Elicitation

Wikis are a collaborative technology in which users can format and post text and images to a Web site. Access control is achieved through password protection and semaphore like protection mechanisms (that is, only one user can write to a particular page of the site at any one time).

Wiki's can be used for collaboration, e.g., to facilitate group work, card entry (for card sorting), template completions, surveys, and for organizing the requirements document. Moreover, Wiki-based requirements can be exported directly to publishing tools and validation tools. In addition, Wikis can be used to build interactive documents that can help to automate test cases (recall that requirements specifications should contain acceptance criteria for all requirements). For example, FitNesse is a free, Wiki-based software collaboration GUI built on top of Fit

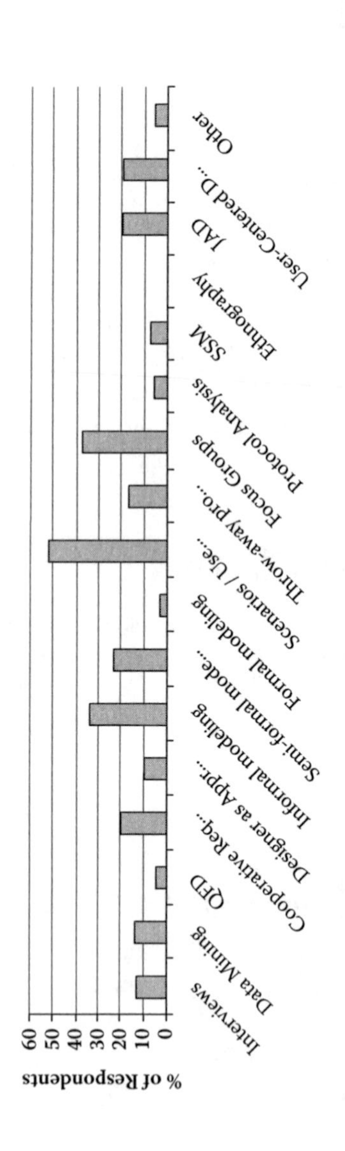

Figure 3.9 Summary of answers to the question "Which requirements elicitation technique(s) do you use? (Neill and Laplante 2003).

framework. FitNesse provides a framework for building Web-based requirements documentation with embedded, interactive test cases (FitNesse 2007). FitNesse is described in further detail in Chapter 8.

Mobile Technologies

Various mobile technologies such as cell phones and personal digital assistants can be used to capture requirements information in situ. For example, while physicians are working with patients, they can transmit information about the activities they are conducting directly to the requirements engineer, without the latter having to be on site. Using mobile devices is particularly useful because they enable instantaneous recording of ideas and discoveries. Such an approach can support brainstorming, scenario generation, surveys, and many other standard requirements elicitation techniques even when the customer is not easily accessible (such as in off-shore software development situations).

A good discussion of the emergence of the use of mobile technologies in requirements discovery can be found in Maiden et al. (2007).

Content Analysis

Content analysis is a technique used in the social sciences for structuring and finding meaning in unstructured information. That is, by analyzing writing or written artifacts of speech, things of importance to the stakeholder can be obtained. In analyzing these writings the objective is to identify recurrent themes. By written artifacts we mean transcripts from group meetings, unstructured interviews, survey data, or emails (any text or artifact that can be converted to text).

Content analysis can be done manually by tagging (colored highlighters) identical words and similar phrases that recur in various writings. For example, Figure 3.10 contains a sample content analysis of some text. The text is an excerpt from a notebook of a requirements engineer who has interviewed a customer regarding a "Smart Home."

As we read the text we begin noticing recurring themes, and as we do so, we highlight them with a different colored marker. For example, the word "I" is mentioned repeatedly, and for whatever reason, we decide that this noun hints at an important theme, and we highlight it in one color. We then notice that "smart home" or "the home" is mentioned several times—we highlight these with the different color from the previous theme. The notion of time is also mentioned frequently ("long periods of time, "time," and "periods of time,") and so we highlight these with a consistent color. And so on.

Free and for-fee tools for content analysis exist to automate this process.

I am gathering requirements for a smart home for a customer

I spend long periods of time interviewing the customer about what she wants

I spend time interacting with the customer as she goes about her day and ask questions like ("why are you running the dishwasher at night, why not in the morning?")

I spend long periods of time passively observing the customer "in action " in their current home to get non-verbal clues about her wants and desires.

I gain other information from the home itself–the books on the book shelf, paintings on the wall, furniture styles, evidence of hobbies, signs of wear and tear on various appliances, etc.

Figure 3.10 Sample content analysis of some random text.

Exercises

3.1. What are some difficulties that may be encountered in attempting to elicit requirements without face-to-face interaction?
3.2. Does the Heisenberg uncertainty principle apply to techniques other than ethnographic observation? What are some of the ways to alleviate the Heisenberg uncertainty principle?
3.3. During ethnographic observation what is the purpose of recording the time and day of the observation made?
3.4. Should requirements account for future scalability and enhancements?
3.5. Which subset of the techniques described in this chapter would be appropriate for a setting where the customers are geographically distributed?
3.6. Investigate the concept of "active listening." How would this technique assist in requirements elicitation?
3.7. If you are working on a class project, what selection of the techniques described in this chapter would you use to elicit system requirements from your customer(s)?

References

Akao, Y. (1990) *Quality Function Deployment: Integrating Customer Requirements into Product Design*, Cambridge, MA: Productivity Press.

Aurum, A., and C. Wohlin (eds) (2005) *Engineering and Managing Software Requirements*, Springer.

Basili, V.R., and D. Weiss (1984) A methodology for collecting valid software engineering data, *IEEE Transactions on Software Engineering*, Nov., pp. 728–738.

The FitNesse project, www.fitnesse.org, last accessed September 15, 2007.

Laplante, P.A. (2006) *What Every Engineer Needs to Know About Software Engineering*, CRC/Taylor & Francis.

Maiden, N., O. Omo, N. Seyff, P. Grunbacher, and K. Mitteregger (2007) Determining stakeholder needs in the workplace: How mobile technologies can help, *IEEE Software*, March/April, pp. 46–52.

Neill, C.J., and P.A. Laplante (2003) Requirements engineering: The state of the practice, *Software*, 20(6): 40–45.

Sommerville, I., and P. Sawyer (1997) Viewpoints for requirements engineering, *Software Quality Journal*, 3: 101–130.

Zowghi, D., and C. Coulin (1998) Requirements elicitation: A survey of techniques, approaches, and tools, in Aurum, A. and C. Wohlin (eds) (2005) *Engineering and Managing Software Requirements*, Springer, pp. 19–46.

Chapter 4

Writing the Requirements Document

Requirements Representation Approaches

Various types of techniques can be used to describe functionality in any system, and the 2003 study that has been repeatedly referenced was intended to discover the prevalence of various specification techniques that were used (Neill and Laplante). Many of the findings with respect to elicitation techniques have been reported already but not with respect to representation approaches.

Generally, there are three approaches to requirements representation: formal, informal, and semi-formal. Requirements specifications can adhere to strictly one or another of these approaches, but usually, they contain elements of at least two of these approaches (informal and one other). Formal representations have a rigorous, mathematical basis, and we will explore these further in Chapter 6. But even formal requirements specifications documents will have elements of informal or semi-formal specification.

Informal requirements representation techniques cannot be completely transliterated into a rigorous mathematical notation. Informal techniques include natural language (that is, human languages), flowcharts, ad hoc diagrams, and most of the elements that you may be used to seeing in systems requirements specifications (SRS). In fact, all SRS documents will have some informal elements. We can state this fact with confidence because even the most formal requirements specification documents have to use natural language, even if it is just to introduce a formal element. We will therefore spend most of the discussion on requirements representation using informal techniques.

Finally, semi-formal representation techniques include those that, while appearing informal, have at least a partial formal basis (e.g., many of the diagrams in the UML family of meta-modeling languages) and those parochial techniques that somehow defy classification as either formal or informal. UML 2.0 is generally considered a semi-formal modeling technique; however, it can be made entirely formal with the addition of appropriate formalisms.

The previously mentioned survey gives some insight into the pervasiveness of formal, informal, and semi-formal approaches to requirements modeling throughout industry, at least within the region of the survey, as shown in Figure 4.1.

Here we see that semi-formal methods (likely UML 2.0) dominate the scene. In the responses to the survey, "other" probably means that the respondent did not understand the nature of the question (Neill and Laplante 2003).

The survey also shed some light on the size of a typical requirements specification, at least in terms of number of requirements in the document (Figure 4.2).

Here it is interesting to see that a relatively large number of requirements specifications (about 20%) were reported to contain between 76 and 150 individual requirements. Only a small number of systems (about 10%) were reported to be "large," having greater than 1200 individual requirements (Neill and Laplante 2003).

Finally, in the same survey we are able to get some sense of the size of the requirements specification document itself (Figure 4.3).

Interestingly we see a relatively high proportion (around 30%) of small (25-50-page) SRS documents, likely corresponding to "small" development projects.

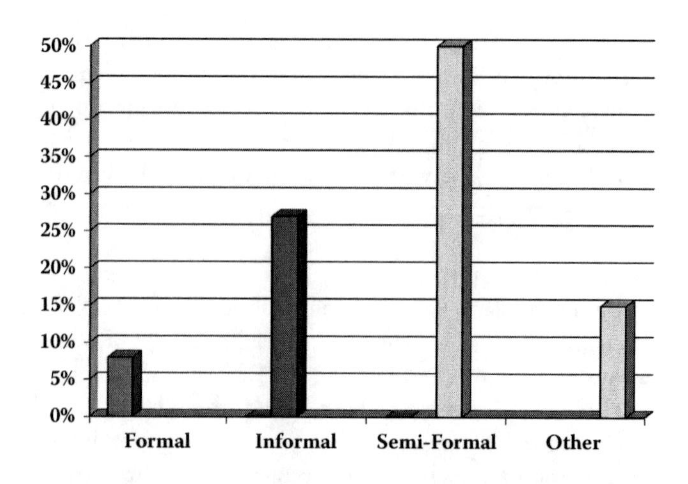

Figure 4.1 Reported requirements notation used (Neill and Laplante 2003).

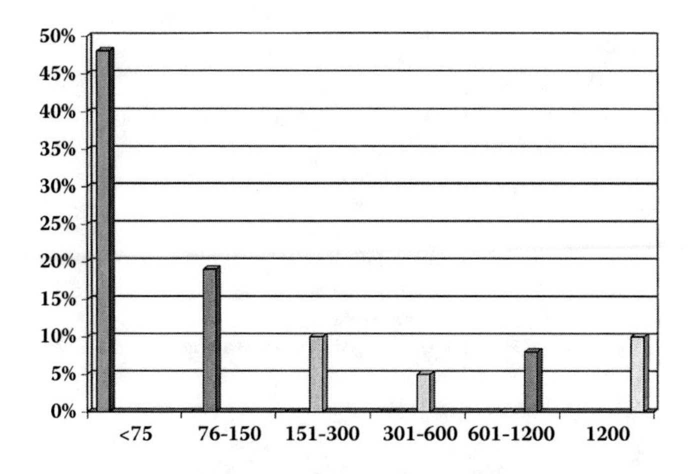

Figure 4.2 Reported number of requirements specifications (Neill and Laplante 2003).

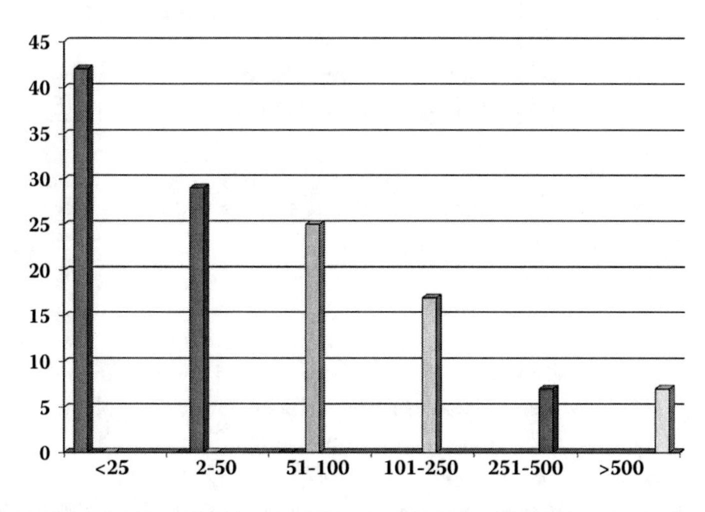

Figure 4.3 Reported number of pages of requirements specifications (Neill and Laplante 2003).

IEEE Standard 830-1998

The IEEE Standard 830-1998, Recommended Practice for Software Requirements Specifications, describes recommended approaches for the specification of software requirements. The standard is based on a model that produces a document that helps

- software customers to accurately describe what they wish to obtain
- software suppliers to understand exactly what the customer wants

1. Introduction
  1.1 Purpose
  1.2 Scope
  1.3 Definitions and Acronyms
  1.4 References
  1.5 Overview
2. Overall description
  2.1 Product perspective
  2.2 Product functions
  2.3 User characteristics
  2.4 Constraints
  2.5 Assumptions and dependencies
3. Specific Requirements
Appendices
Index

Figure 4.4 Table of contents for an SRS document as recommended by IEEE Standard 830.

- individuals to accomplish the following goals:
  - develop a standard software requirements specification (SRS) outline for their own organizations
  - develop the format and content of their specific SRS
  - develop additional supporting items such as an SRS quality checklist or an SRS writer's handbook [IEEE 830]

There are several benefits to observing its guidelines. First, the guidelines provide a simple framework for the organization of the document itself. Moreover, the Standard 830 outline is particularly beneficial to the requirements engineer because it has been widely deployed across a broad range of application domains (Figure 4.4). Finally, and perhaps more importantly, Standard 830 provides guidance for organizing the functional and non-functional requirements of the SRS.

The standard also describes ways to represent functional and non-functional requirements under Section 3, Specific Requirements. These requirements are the next subject for discussion.

IEEE Standard 830 Recommendations on Representing Non-Functional Requirements

Standard 830 defines several types of non-functional requirements involving

- external interfaces
- performance

- logical database considerations
- design constraints
- software system attribute requirements

External interface requirements can be organized in a number of ways, including

- name of item
- description of purpose
- source of input or destination of output
- valid range, accuracy, and/or tolerance
- units of measure
- timing
- relationships to other inputs/outputs
- screen formats/organization
- window formats/organization
- data formats
- command formats

Performance requirements are static and dynamic requirements placed on the software or on human interaction with the software as a whole. Typical performance requirements might include the number of simultaneous users to be supported, the numbers of transactions and tasks, and the amount of data to be processed within certain time periods for both normal and peak workload conditions.

Logical database requirements are types of information used by various functions such as

- frequency of use
- accessing capabilities
- data entities and their relationships
- integrity constraints
- data retention requirements

Design constraint requirements are related to standards compliance and hardware limitations.

Finally, software system attributes can include reliability, availability, security, maintainability, portability, and just about any other "ility" you can imagine.

IEEE Standard 830 Recommendations on Representing Functional Requirements

The functional requirements should capture all system inputs and the exact sequence of operations and responses (outputs) to normal and abnormal situations for every input possibility. Functional requirements may use case-by-case description or other general forms of description (e.g., using universal quantification, use cases, user stories).

The 830 is not prescriptive in terms of how to organize specific functional requirements; instead, a menu of organizational options is offered. Specific functional requirements can be organized by

- functional mode (e.g., "navigation," "combat," "diagnostic")
- user class (e.g., "user," "supervisor," "diagnostic")
- object (by defining classes/objects, attributes, functions/methods, and messages)
- feature (describes what the system provides to the user)
- stimulus (e.g., sensor 1, sensor 2, actuator 1, …)
- functional hierarchy (e.g., using structured analysis)

Or a combination of these techniques can be used within one SRS document.

As an example, the Smart Home description given in the appendix is a feature-driven description of functionality. One clue that it is feature driven is the mantra "the system shall" for many of the requirements.

Alternatively, consider a system organized by functional mode, the NASA WIRE (Wide-field Infrared Explorer) System (1996). This system was part of the "Small Explorer" program. The system describes a submillimeter-wave astronomy satellite incorporating standardized space-to-ground communications. Information was to be transmitted to the ground and commands received from the ground, according to the Consultative Committee for Space Data Systems (CCSDS) and Goddard Space Flight Center standards (NASA 1996).

The flight software requirements are organized as follows:

- system management
- command management
- telemetry management
- payload management
- health and safety management
- software management
- performance requirements

Reviewing the document, under the System Management mode we see the operating system functionality described as follows:

Operating System

001 The operating system shall provide a common set of mechanisms necessary to support real-time systems such as multitasking support, CPU scheduling, basic communication, and memory management.

001.1 The operating system shall provide multi-tasking capabilities.

001.2 The operating system shall provide event-driven, priority-based scheduling.

001.2.1 Task execution shall be controlled by a task's priority and the availability of resources required for its execution.

001.3 The operating system shall provide support for intertask communication and synchronization.

001.4 The operating system shall provide real-time clock support.

001.5 The operating system software shall provide task-level context switching for the 80387 math coprocessor.

Also under system functionality is command validation, defined as follows:

Command Validation

211 The flight software shall perform CCSDS command structure validation.

211.1 The flight software shall implement CCSDS Command Operations Procedure number 1 (COP-1) to validate that CCSDS transfer frames were received correctly and in order.

211.2 The flight software shall support a fixed-size frame-acceptance and reporting mechanism (FARM) sliding window of 127 and a fixed FARM negative edge of 63.

211.3 The flight software shall telemeter status and discard the real-time command packet if any of the following errors occur:

- checksum fails validation prior to being issued;
- an invalid length is detected;
- an invalid Application ID is detected.

211.4 The flight software shall generate and maintain a Command Link Control Word (CLCW). Each time an update is made to the CLCW, a CLCW packet is formatted and routed for possible downlink (NASA 1996).

It is also very common in software-based systems to take an object-oriented approach to describe the system behavior. This is particularly the case when the software is expected to be built using a pure object-oriented language such as Java.

Object-oriented representations involve highly abstract system components called objects and their encapsulated attributes and behavior. The differences between traditional "structured" descriptions of systems and object-oriented descriptions of systems are summarized in Table 4.1.

When dealing with a system organized in an object-oriented fashion, it is very typical to use user stories (especially in conjunction with agile software development methodologies, which we will discuss in Chapter 6) or use cases and use case diagrams to describe behavior.

Table 4.1 Object-Oriented Versus Structured Representation

	Structured	*Object-Oriented*
System components	Functions	Objects
Data and control specification	Separated through internal decomposition	Encapsulated within objects
Characteristics	Hierarchical structure	Inheritance Relationship of objects
	Functional description of system	Behavioral description of system
	Encapsulation of knowledge within functions	Encapsulation of knowledge within objects

ISO/IEC Standard 25030

The recently published ISO/IEC Standard 25030 is designed to be complementary to IEEE Standards 830 and 1223.* ISO 25030 is intended to take a "process view" rather than a product view, with further emphasis on measurable quality requirements. The standard template for 25030, which is similar to that specified by 830, is shown in Figure 4.5.

In particular, Section 6.3 is intended to provide software quality requirements based on measurable targets. This practice is in keeping with IEEE 830's "measurable" characteristics of good requirements. Attribute metrics also assist in comparing values or computing relevant statistics that can lead to system improvement over time (Glinz 2008).

For example, consider a hypothetical requirement 3.4.2 in the baggage handling system:

> 3.4.2 Each baggage scanner unit shall process, on average, 10 pieces of luggage per minute.

A suggested format for the measurable targets based on one suggested in Glinz (2008) is given in Figure 4.6.

Standard 25030 can be used as a template for organizing the SRS document. Alternatively, if using the IEEE 830 or another standard template, Figure 4.6 can also be used to structure metrics and acceptance criteria for requirements, whatever the overall format being used for the SRS.

* IEEE Std Guide for Developing Systems Requirements Specifications, IEEE, 2002. This standard is a software agnostic or systems-oriented version of standard 830.

1. Scope
2. Conformance
3. Normative references
4. Terms and definitions
5. Software quality requirements framework
   - 5.1 Purpose
   - 5.2 Software and systems
   - 5.3 Stakeholders and stakeholder requirements
   - 5.4 Stakeholder requirements and system requirements
   - 5.5 Software quality model
   - 5.6 Software properties
   - 5.7 Software quality measurement model
   - 5.8 Software quality requirements
   - 5.9 System requirements categorization
   - 5.10 Quality requirements life cycle model
6. Requirements for quality requirements
   - 6.1 General requirements and assumptions
   - 6.2 Stakeholder requirements
   - 6.3 System boundaries
   - 6.4 Stakeholder quality requirements
   - 6.5 Validation of stakeholder quality requirements
   - 6.6 Software requirements
   - 6.7 Software boundaries
   - 6.8 Software quality requirements
   - 6.9 Verification of software quality requirements
   - Annex A (Normative). Terms and definitions
   - Annex B (Informative), Processes from ISO/IEC 15288
   - Annex C (Informative), Bibliography

Figure 4.5 Table of contents for requirements specification from ISO/IEC 25030.

- *Attribute:* Average time that a scanner unit needs to scan a piece of luggage
- *Scale:* Seconds (type: ratio scale)
- *Procedure:* Measure time required to scan a package for forbidden contents, take the average over 1000 pieces of various types.
- *Planned value:* 50 percent less than reference value
- *Lowest acceptable value:* 30 percent less than reference value
- *Reference value:* Average time needed by competing or similar products to scan a piece of luggage

Figure 4.6 Sample requirements attribute metrics for baggage handling system.

Standard 25030 is one of five collections of standards known as the "software product quality requirements and evaluation series of standards" (with the acronym "SQUARE"). The other four standards refer to the quality management (ISO/IEC 25000), quality model (ISO/IEC 250100), quality measurement (ISO/IEC 25020), and quality evaluation (ISO/IEC 25040) (Boegh 2008).

Use Cases*

Use cases are an essential element of many SRS documents and are described graphically using any of several techniques. Use cases depict the interactions between the system and the environment around the system, in particular, human users and other systems. One representation for the use case is the use case diagram, which depicts the interactions of the software system with its external environment.

Use cases describe scenarios of operation of the system from the designer's (as opposed to customer's) perspective. Use cases are typically represented using a use case diagram, which depicts the interactions of the software system with its external environment. In a use case diagram, the box represents the system itself. The stick figures represent "actors" that designate external entities that interact with the system. The actors can be humans, other systems, or device inputs. Internal ellipses represent each activity of use for each of the actors (use cases). The solid lines associate actors with each use. Figure 4.7 shows a use case diagram for the baggage inspection system.

Three uses are shown—capturing an image of the baggage ("image baggage"), the detection of a security threat (in which case the bag is rejected from the conveyor for off-line processing), and then configuration by the systems engineer. Notice that the imaging camera, product sensor, and reject mechanism are represented by a human-like stick figure—this is typical—the stick figure represents a system "actor" whether human or not.

Each use case is, however, a form of documentation that describes scenarios of operation of the system under consideration as well as pre- and post-conditions and exceptions. In an iterative development lifecycle these use cases will become increasingly refined and detailed as the analysis and design workflows progress. Interaction diagrams are then created to describe the behaviors defined by each use case. In the first iteration these diagrams depict the system as a "black box," but once domain modeling has been completed the black box is transformed into a collaboration of objects as will be seen later.

Finally, remember that the use case diagram (the picture) is not a use case. It is a visual representation of a use case. To illustrate the difference between a model of

* This discussion is adapted from one found in Laplante (2006), with permission.

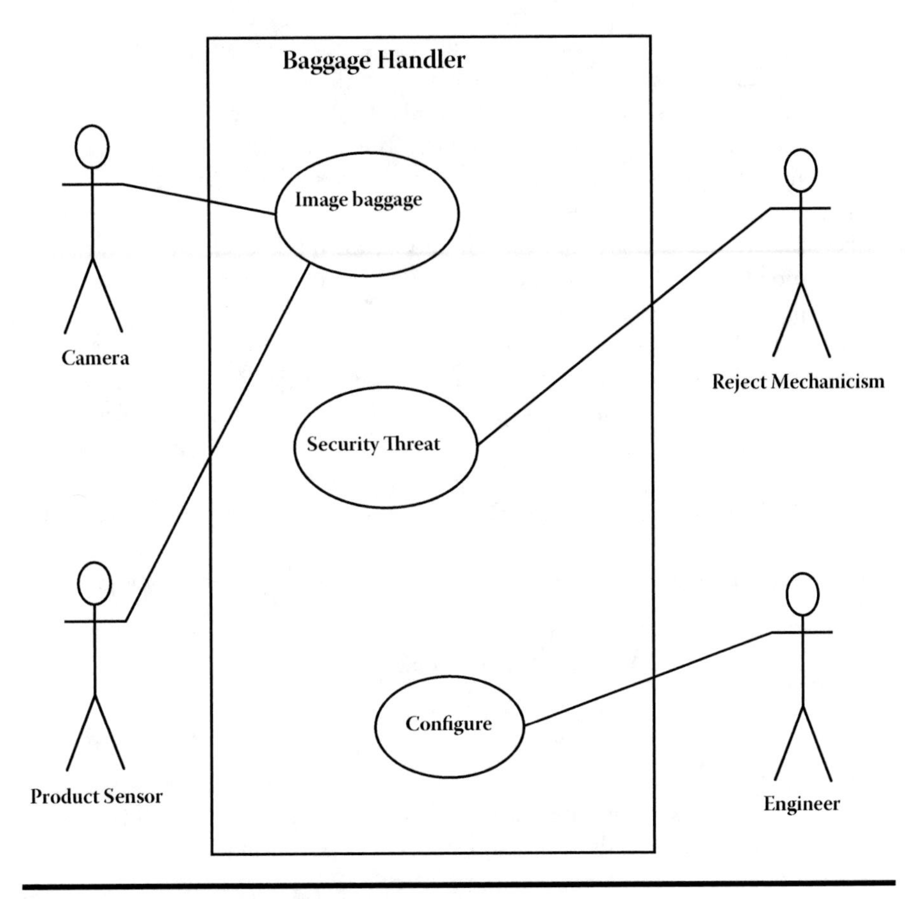

Figure 4.7 Use case diagram of baggage inspection system.

something and that thing, the cover of Craig Larman's book on UML and patterns famously depicts a simple class diagram model of a sailboat with the caption "this is not a sailboat" (Larman 2004).

Behavioral Specifications

In some cases the requirements engineer may be asked to reverse engineer requirements for an existing system when the requirements do not exist, are incomplete, are out of date, or are incorrect. It may also be necessary to generate requirements for open source software (software that is free for use and/or redistribution under the terms of a license) for the purposes of generating test specifications. In these cases a form of SRS, the behavioral specification, needs to be generated.

The behavioral specification is identical in all aspects to the requirements specification, except that the former represents the engineers' best understanding of what the users intended the system to do, while the latter represents the users' best understanding of what the system should do. The behavioral specification has an additional layer of inference and therefore can be expected to be even less complete than a requirements specification.

Fortunately, there is an approach to generating the behavioral specification. The technique involves assembling a collection of as many artifacts as possible that could explain the system's intent. Then, these artifacts are used to reconstruct the system's intended behavior as it is best understood. The forgoing description is adapted from the original paper by Elcock and Laplante (2006).

Artifacts that may be used to derive a behavioral specification include (but should not be limited to)

- any existing requirements specification, even if it is out of date, incomplete, or known to be incorrect
- user manuals and help information (available when running the program)
- release notes
- bug reports and support requests
- application forums
- relevant versions of the software under consideration

Some of these artifacts may have to be scavenged from various sources, such as customer files, emails, open source community repositories, archives, and so forth. A brief description of how these artifacts are used in generating the specification is given below.

Starting with user manuals, statements about the behavior of an application in response to user input can be directly related to the behavioral specification. User manuals are particularly well suited for this purpose since describing the response of software to user stimulus is germane to their existence. Help information, such as support Web sites and application help menus, are also rich in information that can either be used directly or abstracted to define behavioral requirements.

Next, release notes can be consulted. Release notes are typically of limited benefit to the process of developing test cases since they tend to focus on describing which features are implemented in a given release. There is usually no description of how those supported features should function. However, release notes are particularly important in resolving the conflict that arises when an application does not respond as expected for features that were partially implemented in or removed from future implementation.

Defect reports can also be a great source for extracting behavioral responses because they are typically written by users, they identify troublesome areas of the software, and they often clarify a developer's intention for the particular behavior. Defect reports, at least for open source systems, are readily found in open repositories such as Bugzilla. While it is true that in some cases bug reports contain too much implementation detail to be useful for extracting behavioral responses, they can be discarded if behavior cannot be extrapolated from the details.

In many ways, the content of support requests is similar to bug reports in that they identify unexpected behavior. Unlike bug reports, however, support requests can sometimes be helpful in identifying features that are not fully implemented as well as providing information that illuminates the expectations of users. It is important to investigate both since, in addition to providing the insights mentioned, they may also aid in rationalizing unexpected behavior.

For many open source projects, and some closed source projects, there are Web-based forums associated with the project. Within these forums, various amounts of behavioral information can be extracted. Ranging from useless to relevant, open-discussion postings need to be carefully filtered and applied only when other development artifacts are lacking. As with other artifacts, these postings can also be used to clarify behavior.

Finally, in the absence of any other artifacts, the software being tested could itself be an input to developing structural (glass-box) test cases. Assuming that this approach is necessary, the reality is that it will really be the tester's characterization of correct behavior that will largely prevail in defining the test cases.

Once the discovery process has concluded, the behavioral specification can be written. The format of the behavioral specification is identical to that for requirements specification, and all of the IEEE 830 rules should be applied (Elcock and Laplante 2006).

The Requirements Document

The system (or software) requirements specification document (SRS) is the official statement of what is required of the system developers. It is important to remember that, under those circumstances where there is a customer-vendor relationship between the sponsor and builders of the system, the SRS is a contract and is therefore enforceable under civil contract law (or criminal law if certain types of fraud or negligence can be demonstrated).

The author is often asked "what is the correct format for an SRS?" In many cases students want templates, or they are literally interested in the size of type font, margin dimensions, and so forth. But there are many documentation formats, and none is better than any other (except, of course, in the case of a hard-to-read, badly organized, and imprecise document). The "right" format all depends on what the sponsor, situation, customer, application domain, your employer, and so forth demand.

Whatever the format, it is important to remember that the SRS is NOT a design document. As far as possible, it should set forth WHAT the system should do rather than HOW it should do it.

Users of a Requirements Document

There are several "users" of the requirements document, each with a unique perspective, needs, and concerns. Typical users of the document include

- customers
- managers
- developers
- test engineers
- maintenance engineers
- stakeholders

Customers specify the requirements and are supposed to review them to ensure that they meet their needs. Customers also specify changes to the requirements. Because customers are involved, and they are likely not engineers, SRS documents should be accessible to the lay person (formal methodologies excepted).

Managers at all levels will use the requirements document to plan a bid for the system and to plan for managing the system development process. Managers, therefore, are looking for strong indicators of cost and time-to-complete in the SRS.

And of course, developers use the requirements specification document to understand what system is to be developed. At the same time, test engineers use the requirements to develop validation tests for the system. Later, maintenance engineers will use the requirements to help understand the system and the relationship between its parts so that the system can be upgraded or fixed. Other stakeholders who will use the SRS include all of the direct and indirect beneficiaries (or adversaries) of the system in question, as well as lawyers, judges, plaintiffs, juries, district attorneys, arbiters, mediators, etc., who will view the SRS as a legal document in the event of disputes.

Requirements Document Requirements

That the SRS document should be easy to change is evident for the many reasons we have discussed so far. Furthermore, since the SRS document serves as a reference tool for maintenance, it should record forethought about the lifecycle of the system, that is, to predict changes.

There is no "silver bullet" for a format for the document—each system should be considered on its own merits. But in terms of general organization, writing approach, and discourse, best practices include

- using consistent modeling approaches and techniques throughout the specification, for example, a top-down decomposition, structured, or object-oriented approaches
- separating operational specification from descriptive behavior

- using consistent levels of abstraction within models and conformance between levels of refinement across models
- modeling nonfunctional requirements as a part of the specification models—in particular timing properties
- omitting hardware and software assignments in the specification (another aspect of design rather than specification)

Following these rules will always lead to a better SRS document.

Preferred Writing Style

Engineers (of all types) have acquired an unfair reputation for poor communications skills, particularly writing. In any case, it should be clear now that requirements documents should be very well written. It is not appropriate for us to offer recommendations on writing here. But we urge you to improve your writing through practice, study (of writing techniques), and through reading—yes, read "well-written" SRS documents to learn from them. You should also read literature, poetry, and news, as much can be learned about economy and clarity of presentation from these writings. Some have even suggested screenwriting as an appropriate paradigm for writing requirements documents (or for user stories and use cases) (Norden 2007).

In any case approach the requirements document like any writing—be prepared to write and rewrite, again and again. Have the requirements document reviewed by several other stakeholders (and possibly a non-stakeholder who writes well). Metrics can be helpful in policing basic writing features (for example, average word, sentence, and paragraph length). Most word processing tools calculate these metrics for you, and certain metrics will be discussed in Chapter 5. But we would like to discuss structural metrics now.

Text Structure

Numbering structure depth is a metric that counts the numbered statements at each level of the source document. For example, first-level requirements, numbered 1.0, 2.0, 3.0, and so forth, are expected to be very high-level (abstract) requirements. Second-level requirements numbered 1.1, 1.2, 1.3, …, 2.1, 2.2, 2.3, …, 3.1, 3.2, etc. are subordinate requirements at a lower level of detail. Even more detailed requirements will be found at the third level, numbered as 1.1.1, 1.1.2, and so on. A specification can continue to fourth or even fifth-level requirements, but normally, third or fourth levels of detail should be sufficient. In any case, the counts of requirements at level 1, 2, 3, and so on provide an indication of the document's organization, consistency, and level of detail.

A thoughtful and well-organized SRS document should have a consistent level of detail, and if you were to list out the requirements at each level, the resultant

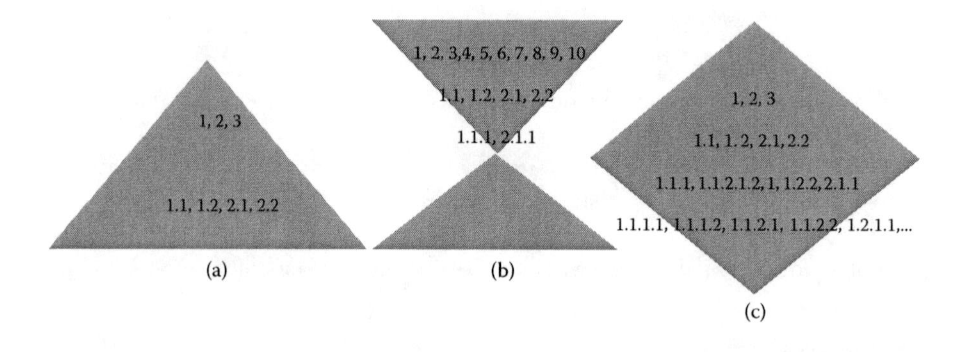

Figure 4.8 Pyramid, hourglass, and diamond-shaped configurations for requirements text structure.

shape should look like a pyramid in that there should be a few numbered statements at level 1 and each lower level should have increasingly more numbered statements than the level above it (Figure 4.8a).

On the other hand, requirements documents whose requirements counts at each level resemble an hour-glass shape (Figure 4.8b) are usually those that contain a large amount of introductory and administrative information. Finally, diamond-shaped documents, represented by a pyramid followed by decreasing statement counts at lower levels (Figure 4.8c) indicate an inconsistent level of detail representation (Rosenberg et al. 2008).

The NASA ARM tool introduced in Chapter 5 computes text structure for a given SRS document in an appropriate format.

Best Practices and Recommendations

Writing effective requirements specifications can be very difficult even for trivial systems because of the many challenges that we have noted. Some of the more common dangers in writing poor SRS documents include

- mixing of operational and descriptive specifications
- combining low-level hardware functionality and high-level systems and software functionality in the same functional level
- omission of timing information

Other bad practices arise from failing to use language that can be verified. For example, consider this set of "requirements":

- The system shall be completely reliable.
- The system shall be modular.

- The system will be fast.
- Errors shall be less than 99%.

What is wrong with these? The problem is they are completely vague and immeasurable and therefore unachievable. For example, what does "completely reliable" mean? Any arbitrary person will have a different meaning for reliability for a given system. Modularity (in software) has a specific meaning, but how is it measurable? What does "fast" mean? Fast as a train? Faster than a speeding bullet? This requirement is just so vague. And finally "errors shall be less than 99%" is a recipe for a lawsuit. 99% of what?, over what period of time?, and so forth.

For the above set of requirements a slightly better version might be

- Response times for all level one actions will be less than 100 ms.
- The cyclomatic complexity of each module shall be in the range of 10 to 40.
- 95% of the transactions shall be processed in less than 1 s.
- Mean time before first failure shall be 100 hours of continuous operation.

But even this set is imperfect, because there may be some important details missing—we really can't know what, if anything, is missing outside the context of the rest of the SRS document.

Some final recommendations for the writing of specification documents are

- Invent and use a standard format and use it for all requirements.
- Use language in a consistent way.
- Use "shall" for mandatory requirements.
- Use "should" for desirable requirements.
- Use text highlighting to identify key parts of the requirement.
- Avoid the use of technical language unless it is warranted.

But we are not done yet with the writing of requirements specifications. To this goal, Chapter 5 is devoted to perfecting the writing of specific requirements.

Exercises

4.1 Under what circumstances is it appropriate to represent an SRS using informal techniques only?

4.2 What can the behavioral specification provide that a requirements document cannot?

4.3 If the customer requests that future growth and enhancement ideas be kept, where can these ideas be placed?

4.4 What are some items to be included under "data retention" in the SRS?

4.5 Here are some more examples of vague and ambiguous requirements that have actually appeared in real requirements specifications. Discuss why they are vague, incomplete, or ambiguous. Provide improved versions of these requirements (make necessary assumptions).

4.5.1 The tool will allow for expedited data entry of important data fields needed for subsequent reports.

4.5.2 The system will provide an effective means for identifying and eliminating undesirable failure modes and/or performance degradation below acceptable limits.

4.5.3 The database creates an automated incident report that immediately alerts the necessary individuals.

4.5.4 The engineer shall manage the system activity, including the database.

4.5.5 The report will consist of data, in sufficient quantity and detail, to meet the requirements.

4.5.6 The data provided will allow for a sound determination of the events and conditions that precipitated the failure.

4.5.7 The documented analysis report will include, as appropriate, investigation findings, engineering analysis, and laboratory analysis.

4.6 In Section 9.4 of the SRS in the appendix, which requirements are suitable for representing using the "measurable targets" in the format shown in Figure 4.6?

4.7 Many of the requirements in the appendix can be improved in various ways. Select ten requirements listed and rewrite them in an improved form. Discuss why your rewritten requirements are superior using the vocabulary of IEEE 830.

References

Boegh, J. (2008) A new standard for quality requirements, *Computer*, 25(2): 57–63.

Elcock, A., and P.A. Laplante (2006) Testing without requirements, *Innovations in Systems and Software Engineering: A NASA Journal*, 2: 137–145.

Glinz, M. (2008) A risk-based, value-oriented approach to quality requirements, *Computer*, 25(2): 34–41.

IEEE Standard 830-1998, *Recommend Practice for Software Requirements Specifications*, IEEE Standards Press, Piscataway, NJ, 1998.

Laplante, P.A. (2006) *What Every Engineer Needs to Know About Software Engineering*, CRC/Taylor & Francis.

Larman, C. (2004) *Applying UML and Patterns*, 3rd Edition, Prentice-Hall PTR.

NASA WIRE (Wide-Field Infrared Explorer) Software Requirements Specification (1996) http://sunland.gsfc.nasa.gov/smex/wire/mission/cdhsw/wirrqtop.htm, last accessed 3 February 2008.

Neill, C.J., and P.A. Laplante (2003) Requirements engineering: The state of the practice, *Software*, 20(6): 40–45.

Norden, B. (2007) Screenwriting for requirements engineers, *Software*, 26–27.

Rosenberg, L., T. Hammer, and L. Huffman, Requirements, Testing, & Metrics, NASA Software Assurance Technology Center Report, http://satc.gsfc.nasa.gov, last accessed 2 February 2008.

Chapter 5

Requirements Risk Management

What Is Requirements Risk Management?

To motivate the notion of requirements risk, here is a vignette.* On a road that the author travels, he passes a strange street sign that declares "End brake retarder prohibition." The meaning of this sign is hard to understand, and is blurred by the curious use of a quadruple negative (each word in the directive has the connotation of stopping something). As it turns out, this sign is an exquisite example of a "shall not" requirement as well as illustrating a poor requirements specification, namely, one that is ambiguous, vague, contradictory, incomplete, or contains a mixed level of abstraction.

What is a "brake retarder"? Briefly, it is a device used on large trucks to slow the engine down by allowing air to be exhausted out of the pistons, thus slowing the vehicle. Brake retarders are very noisy, and, as a result, many municipalities prohibit their use within city limits. By Pennsylvania law, appropriate signs must be posted with the instructions "Brake Retarders Prohibited Within Municipal Limits." The complementary sign then reads "End Brake Retarder Prohibition." The latter phrasing is, apparently, unique to the Commonwealth of Pennsylvania (O'Neil 2004). In some states the sign pair reads "No Jake Brakes" and "End

* A variation of this story first appeared in "End Brake Retarder Prohibitions: Defining 'Shall Not' Requirements Effectively," *Computer*, 2009, by Jeff Voas and Phil Laplante.

Jake Brake Prohibition," the term "Jake" being a nick-name of a company that manufacturers the device, Jacobs Vehicle Systems.

Taking the Pennsylvania form of the sign, if we substitute appropriate synonyms:

end → stop, brake → stop, retarder → stopper, prohibition → stopping

the sign translates to:

"stop stop stopper stopping"

Which is a quadruple negative. In theory, we should be able to replace such a quadruple negative with a null sign reading:

Ø

The original sign has the paradoxical effect, however, of causing a positive action (that is, to allow the application of brake retarders).

It is clear that the "brake retarder" sign is a poor one—but perhaps not nearly as poor as the requirements for installing one. Pennsylvania's requirements on municipalities who wish to implement a "no brake retarder" zone are

1. Downhill grade(s) greater than 4%
2. A posted reduced speed limit for trucks due to a hazardous grade determination
3. Posted reduced gear zone(s)
4. Posted speed limits over 55 miles per hour
5. Highway exit ramps with a posted speed limit over 55 miles per hour

and the crash history for the stretch of road a municipality is seeking to keep brake-retarder free *must not include*

1. *A history of runaway truck crashes over the past three years*
2. *A discernible pattern of rear-end crashes over the past three years where the truck was the striking vehicle* (O'Neil 2004)

The italic font was added to the latter section because it highlights the fact that the requirements on brake retarder prohibitions contain an embedded shall not requirement.

Given all of this confusion, clearly, a better job in formulating the sign language is needed, and more work is needed in formulating the requirements for the sign. Formal methods may have been advisable in this regard. All of these activities fall under the context of requirements risk management.

Requirements risk management involves the proactive identification, monitoring, and mitigation of any factors that can threaten the integrity of the requirements engineering process. Requirements risk factors can be divided into two types: technical and management. Technical risk factors pertaining to the elicitation, agreement, and representation processes have already been discussed in Chapter 3. In Chapter 6 we discuss the use of formal methods in improving the quality of requirements representation through mathematically rigorous approaches. Requirements management risk factors tend toward issues of expectation management and interpersonal relationships, and these are discussed in Chapter 9. In this chapter we wish to focus on the mitigation of requirements risk through the analysis of the requirements specification document itself. That is, the validation and verification of the SRS occur early in order to avoid costly problems further downstream. There are a variety of complementary and overlapping techniques to check quality attributes of the SRS, but IEEE Standard 830 contains a set of rules that are extremely helpful in vetting the technical aspects of the SRS and, in turn, mitigating risk later on. So in this chapter we will be looking at the nature of SRS verification and validation, at various qualities of "goodness" for requirements specifications, and finally, turn to work at NASA that can be very helpful in comprehending these qualities of goodness.

Requirements Validation and Verification

Requirements validation and verification involves review, analysis, and testing to ensure that a system complies with its requirements. Compliance pertains to both functional and nonfunctional requirements. But the foregoing definition does not readily distinguish between verification and validation and is probably too long in any case to be inspirational or to serve as a mission statement for the requirements engineer. Barry Boehm (1984) suggests the following to make the distinction between verification and validation:

requirements validation: "am I building the right product?"
requirements verification: "am I building the product right?"

In other words, validation involves fully understanding customer intent and verification involves satisfying customer intent.

There are terrific benefits to implementing a requirements verification and validation program. These include

- early detection and correction of system anomalies
- enhanced management insight into process and product risk
- support for lifecycle processes to ensure conformance to program performance and budget
- early assessment of software and system performance

- ability to obtain objective evidence of software and system conformance to support process
- improved system development and maintenance processes
- improved and integrated systems analysis model (IEEE 1012)

Requirements validation involves checking that the system provides all of the functions that best support the customer's needs and that it does not provide the functions that the customer does not need or want. Additionally, validation should ensure that there are no requirements conflicts and that satisfaction of the requirements can actually be demonstrated. Finally, there is some element of sanity check in terms of time and budget—a requirement may be able to be literally met, but the cost and time needed to meet the requirement may be unacceptable or even impossible.

Requirements verification (testing) involves checking satisfaction of a number of desirable properties of the requirements (e.g., IEEE 830 rules). Usually we do both validation and verification (V&V) simultaneously, and often the techniques used for one or the other are the same.

Techniques for Requirements V&V

V&V techniques may include some of the requirements elicitation techniques that were discussed in Chapter 3. For example, group reviews/inspections, focus groups, prototyping, viewpoint resolution, or task analysis (through user stories and use cases) can be used to simplify, combine, or eliminate requirements. In addition, we can use comparative product evaluations to uncover missing or unreasonable requirements and task analysis to uncover and simplify requirements. Systematic manual analysis of the requirements, test-case generation (for testability and completeness), using an executable model of the system to check requirements, and automated consistency analysis may also be used. Finally, certain formal methods such as model checking and consistency analysis can be used for V&V and these will be discussed in Chapter 6.

In any case we now present some examples of various techniques for requirements verification and validation.

Goal-Based Requirements Analysis

Stakeholders tend to express their requirements in terms of operations and actions rather than goals. A risk is posed when goals evolve as stakeholders change their minds and refine and operationalize goals into behavioral requirements. To reduce this risk stakeholder goals need to be evolved until they can be structured as requirements.

Goal evolution is facilitated through goal elaboration and refinement. Useful techniques for goal elaborating include identifying goal obstacles, analyzing scenarios and constraints, and operationalizing goals. Goal refinement occurs when synonymous goals are reconciled, when goals are merged into a subgoal categorization, when constraints are identified, and when goals are operationalized (Hetzel 1988). As to operationalization, we'll refer to goals-based analysis when discussing metrics generation.

Requirements Understanding

In an early and influential book on software testing, Bill Hetzel proposed several paradigms for requirements verification and validation. To set the stage for this V&V, Hetzel addressed the problem of requirements understanding through the following analogy.

Imagine you are having a conversation with a "customer" in which he says that he would like for you to develop some kind of health management system, which, among other things, ensures that patients are eating a "well-balanced meal." You readily agree to this requirement. Many weeks later as you begin thinking about a system design, you reconsider the requirement to provide a "well-balanced meal." What does that mean? In one interpretation, it could mean, adding up everything consumed; were minimum nutritional guidelines in terms of calories, protein, vitamins, and so forth met? Another interpretation is that the patient ate exactly three meals. Yet another interpretation is that the meals were "well balanced" in the sense that each food item weighed the same amount (Hetzel 1988).

Aside from the more ridiculous interpretations of "well-balanced meal," the previous example illustrates a problem. "Well-balanced meal" may have no language equivalent in French, Hindi, Mandarin, or any other language. So "well-balanced meal" would create a problem for any non-native English speaker. What other colloquialisms do we use in our specifications and then ship out for offshore development? Clearly, there are various problems that can arise from language and cultural differences.

One solution to the requirements understanding problem is offered by Hetzel. He suggests that for correct problem definition it is best to specify the test for accepting a solution along with the statement of the requirement. When the statement and test are listed together, most problems associated with misunderstanding requirements disappear. In particular we want to derive requirements-based test situations and use them as a test of requirement understanding and validation.

For example, when a requirement is found to be incomplete, we can use the test case to focus on missing information. That is, design the test case to ask the question, "What should the system do in this case when this input is not supplied?" Similarly, when a requirement is found to be fuzzy or imprecise, use a test case to ask the question, "Is this the result I should have in this situation?" The specific instance will focus attention on the imprecise answer or result and ensure that it is examined carefully (Hetzel 1988). Today, Hetzel's approach is called "test-driven development."

Validating Requirements Use Cases

When use cases comprise part of the requirements, these can be validated by asking a simple set of questions:

- Are there any additional actors that are not represented?
- Are there any activities that are not represented?
- Are each actor's goals being met?
- Are there events in the use case that do not address these goals?
- Can the use case be simplified?

Other related questions can and should be readily generated.

Prototyping

Prototypes are useful in V&V when very little is understood about the requirements or when it is necessary to gain some experience with the working model in order to discover requirements. The principle behind using working prototypes is the recognition that requirements change with experience and prototyping yields that experience.

There are two kinds of prototypes—throwaway and non-throwaway. Throwaway prototypes are designed to be quickly built and then discarded. Non-throwaway prototypes are intended to be used as the basis for the working code. The advantage of throwaway code is that it can be more quickly built. But non-throwaway code does not waste effort. Of course, care should be taken to ensure that throwaway code does not end up being kept because it was likely not built robustly.

Incremental and evolutionary development approaches are essentially based on a series of non-throwaway prototypes. The difference between the two approaches is, essentially, that in incremental development the functionality of each release is planned, whereas in evolutionary development, subsequent releases are not planned out. In both incremental and evolutionary development, lessons learned from prior releases inform the functionality of future releases' incremental and evolutionary development. In essence, early versions are prototypes used for future requirements discovery.

The Requirements Validation Matrix

A requirements validation matrix is an artifact that connects requirements to the tests associated with that requirement. Such a matrix facilitates review of requirements and the tests and provides an easy mechanism to track the status of test case design and implementation. An excerpt from a requirements validation matrix for the Smart Home system SRS documents is shown in Table 5.1.

Table 5.1 A Sample Requirements Validation Matrix for the Smart Home System SRS in the Appendix

Requirement	*Test Cases*	*Status*
9.13.1 System shall provide wireless support for driving any number of wall mounted monitors for picture display.	T-1711 T-1712 T-1715	Passed Passed Passed
9.13.2 System shall provide Web-based interface for authenticated users to publish new photos for display on wall monitors.	T-1711 T-1715 T-1811	Passed Failed Passed
9.13.3 System shall allow users to configure which pictures get displayed.	T-1712 T-1715 T-1811 T-1812 T-1819	Passed Passed Passed Passed Not run
9.13.4 System shall allow users to configure which remote users can submit pictures to which wall monitor.	T-1712 T-1715 T-1716 T-1812	Passed Passed Passed Passed

Here the requirements forming the SRS are listed verbatim in the far left column, the tests that verify those requirements listed in the center column, and the test case status in the far right column. The status field can contain "passed," "failed," "not run," "omitted," or any variation of these keywords. Additional columns can be added to the matrix to indicate when the test was run, who conducted the test, where the test was run, the status of testing equipment used, and for comments and other relevant information.

The requirements validation matrix is easily made part of the master test plan and can be updated throughout the project to give a record of all requirements testing.

The Importance of Measurement in Requirements Verification and Validation

Imagine an argument involving the question of which boxer was better: Muhammad Ali or Joe Louis? Now, all kinds of elaborate simulation models have been contrived to settle this argument. You can try to argue superiority based on particular fighting characteristics. Some have modeled these characteristics to create simulated fights. Still others have looked at the fighting characteristics and results against similar opponents. You could even argue about training techniques and the quality of their managers, trainers, and promoters. But there is no conclusive evidence that one fighter was better than the other or that they were evenly matched because there are no direct metrics available to make just a value judgment. Now consider an argument involving who is the best long jumper of all time. The answer should be easy—Mike Powell, who holds the record at 29 feet, 4 and 3/8 inches (with no wind at his back). No one has ever jumped farther, and he broke Bob Beamon's record, which stood for 23 years. You can try to argue that Beamon was better in terms of other characteristics—competitive spirit, tenacity, resiliency, sportsmanship, etc., but those are all unmeasurable.

Now imagine an argument with a customer involving whether or not a requirement that "the software shall be easy to use" was met or not. You contend that the requirement was met because "look, the software is easy to use." The customer disagrees because she feels that "the software is too hard to use." But you can't win the argument because you have no metrics. It is rather disappointing that as software engineers we are often no better off than two boxing pundits arguing from barstools (Laplante et al. 2007).

So which qualities should you consider and measure? Any collection of qualities is sometimes referred to as "the ilities." There are many possible qualities that comprise the "ilities" including:

- accuracy
- completeness
- consistency
- correctness
- efficiency
- expandability
- interoperability
- maintainability
- manageability
- portability
- readability
- reusability
- reliability
- safety
- security
- survivability
- testability
- understandability
- usability

This is not an exhaustive list. In any case, for each requirement containing one of the ilities there needs to be an associated metric to determine if the requirement has been met. Requirements measurement is a concept that we will be revisiting over and over again.

Goal/Question/Metric Analysis

We previously mentioned the use of goal-based analysis for requirements verification and validation. But how can we generate the metrics that we need?

The goal/question/metric (GQM) paradigm is an analysis technique that helps in the selection of an appropriate metric. To use the technique, you follow three simple rules. First state the goals of the measurement, that is, "what is the organization trying to achieve?" Next, derive from each goal the questions that must be answered to determine if the goals are being met. Finally, decide what must be measured in order to be able to answer the questions (Basili et al. 1994).

Here is an example of using GQM to define metrics that are "useful." Suppose that the stated goal for the system is "The system shall be easy to use." Hopefully, you will agree that "easy to use" is impossible to objectively measure. So how do we approach its measurement? We do it by creating questions that help describe what "easy to use" means. For example, one question that fits this description is "How many expert, intermediate, and novice users use the system?" The rationale for this question is that an easy-to-use system should be used by everyone. Now we need to know an appropriate metric to answer this question. Here is one way to obtain that metric—provide the system in an open lab for a period of time and measure the number and percentage of each user type who uses the system during that time. If a disproportionate number of users are expert, for example, then it may be concluded that the system is not easy to use. If an equal proportion of expert, intermediate, and novice users use the system, then it might be that the system is "easy to use."

Consider another question that addresses the goal of "easy to use": How long does it take a new user to master features 1 through 25 with only 8 hours of training? The rationale is that certain minimum features needed to use the system adequately ought not to take too long to train. An associated metric for this question then is obtained by taking a random sample of novice users, give them the same 8 hours of training, and then testing the students to see if they can use features 1–25 to some minimum standard.

Following such a process to drive questions and associated metrics from goals is a good path to deriving measureable requirements and at the same time helping to refine and improve the quality of the requirements themselves.

Standards for Verification and Validation

There are various international standards for the processes and documentation involved in verification and validation of systems and software. Many of these have been sponsored or co-sponsored by the Institute for Electrical and Electronics Engineers (IEEE).

Whatever requirements V&V techniques are used, a software requirements V&V plan should always be written to accompany any major or critical software application.

IEEE Std 1012-2004, IEEE Standard for Software Verification and Validation, provides some guidelines to help prepare verification and validation plans. Figure 5.1 shows the recommended V&V plan outline (IEEE 1012).

1. Purpose
2. Referenced documents
3. Definitions
4. V&V overview
    - 4.1 Organization
    - 4.2 Master schedule
    - 4.3 Software integrity level scheme
    - 4.4 Resources summary
    - 4.5 Responsibilities
    - 4.6 Tools, techniques, and methods
5. V&V processes
    - 5.1 Process: Management
    - 5.1.1 Activity: Management of V&V
    - 5.2 Process: Acquisition
        - 5.2.1 Activity: Acquisition support V&V
    - 5.3 Process: Supply
        - 5.3.1 Activity: Planning V&V
    - 5.4 Process: Development
        - 5.4.1 Activity: Concept V&V
        - 5.4.2 Activity: Requirements V&V
        - 5.4.3 Activity: Design V&V
        - 5.4.4 Activity: Implementation V&V
        - 5.4.5 Activity: Test V&V
        - 5.4.6 Activity: Installation and checkout V&V
    - 5.5 Process: Operation
        - 5.5.1 Activity: Operation V&V
    - 5.6 Process: Maintenance
        - 5.6.1 Activity: Maintenance V&V
6. V&V reporting requirements
    - 6.1 Task reports
    - 6.2 Activity summary reports
    - 6.3 Anomaly reports
    - 6.4 V&V final report
    - 6.5 Special studies reports (optional)
    - 6.6 Other reports (optional)
7. V&V Administrative requirements
    - 7.1 Anomaly resolution and reporting
    - 7.2 Task iteration policy
    - 7.3 Deviation policy
    - 7.4 Control procedures
    - 7.5 Standards, practices, and conventions
8. V&V test documentation requirements

Figure 5.1 Recommended V&V plan table of contents (IEEE 1012).

IEEE Standard 830

IEEE 830 is perhaps the most important of all the standards that relate to requirements engineering. IEEE 830 "describes recommended approaches for the specification of software requirements." The standard attempts to help

a) Software customers to accurately describe what they wish to obtain;
b) Software suppliers to understand exactly what the customer wants;
c) Individuals to accomplish the following goals:

1) Develop a standard software requirements specification (SRS) outline for their own organizations;
2) Define the format and content of their specific software requirements specifications;
3) Develop additional local supporting items, such as an SRS quality checklist or an SRS writer's handbook." (IEEE 830)

But from a risk mitigation standpoint, we are most interested in the qualities of goodness for requirements document that are described. These are

- correct
- unambiguous
- complete
- consistent
- ranked for importance and/or stability
- verifiable
- modifiable
- traceable

Let us describe each of these qualities in some detail.

Correctness

Correctness means that any requirement listed is one that needs to be met (i.e., incorrect requirements specifications specify unwanted behavior). Correctness is an important quality of an SRS document because unwanted behavior is, well, unwanted.

Sometimes unwanted behavior is not always obvious. Consider the following requirement for a computer security system.

2.1.1 All passwords and user ids shall be unique.

There is a problem with this requirement, for, if user A tries to set a password that is already taken by another user, the system has to indicate so. This gives user A knowledge of the password of another user, which could be used for an attack. Clearly this is incorrect behavior that we do not want to specify.

Various techniques can be used to study correctness including reviews and inspections, but none of these is perfect and it is probably the case that more than one review and/or inspection needs to be employed to ensure correctness.

Ambiguity

We define ambiguity by complementation—an SRS document is unambiguous if each specification element can have only one interpretation.

Here is an example of a specification document in which ambiguous behavior was described. In a certain automobile (belonging to the author) an indicator light is displayed (in the shape of an engine) when certain exceptional conditions occur. These conditions include poor fuel quality, fuel cap not tightened properly, and other fuel-related faults. According to the user's manual, if the cause of the problem is relatively minor, such as the fuel cap not tightened, the system will reset the light upon:

> removing the exceptional condition followed by three consecutive error free cold starts. A cold start is defined as a start up that has not been preceded by another engine start up in the last eight hours, followed by several minutes of either highway or city driving.*

Can you see the problem in this functional definition? Aside from its confusing wording, the requirement doesn't make sense. If you wait eight hours from the previous start up, then start the engine to drive somewhere, you have to wait at least eight hours to start up and drive back to your origin. If you have any warm start before three consecutive cold starts, the sequence has to begin again. Is the only possible way to satisfy this condition to drive somewhere, wait there for eight hours and then drive back to the origin (three times in a row)? Or, drive around for a while, return to the origin, wait eight hours, then do it again (two times more)? This sequence of events is very hard to follow, and in fact, after one month of trying, the author could not get the light to reset without disconnecting and reconnecting the battery.

Another reason why ambiguity is so dangerous is that, in an ambiguous requirements specification, literal requirements satisfaction may be achieved but not customer satisfaction. "I know that is what I said I wanted, but now that I see it in action, I realized that I really meant something else" is an unfortunate refrain. Or consider this fictitious quote—"oh, you meant THAT lever; I thought you meant the other one." We would never want this scenario to be played out late in the system's life cycle.

Some of the techniques that could be used to resolve ambiguity of SRS documents include formal reviews, viewpoint resolution, and formal modeling of the specification.

Completeness

An SRS document is complete if there is no missing functionality, that is, all appropriate desirable and undesirable behaviors are specified. Recall from Figure 1.4 that there is usually a mismatch between desired behaviors and specified behaviors—there is always some unspecified behavior, as well as undesirable behavior, that finds

* This is not a verbatim quote. It is a representation of the manual's verbiage to avoid exposing the identity of the vehicle.

its way into the system that should be explicitly prohibited. Either case can lead to literal requirements satisfaction but not customer satisfaction.

Completeness is a difficult quality to improve. How do you know when something is missing? Typical techniques for reducing incompleteness include various reviews, viewpoint resolution, and the act of test case generation. Test-driven development, which will be discussed in Chapter 7, has the effect of asking "what's missing from here" and "what can go wrong?" Answering these questions will tend to lead to systems features being uncovered. QFD is also a powerful technique to combat incompleteness because of the comparison of the system under consideration with competing systems.

Consistency

The consistency of the SRS document can take two forms: internal consistency—i.e., satisfaction of one requirement does not preclude satisfaction of another; and external consistency—i.e., the SRS is in agreement with all other applicable documents and standards.

When either internal or external inconsistency is present in the SRS, it can lead to difficulties in meeting requirements and delays and frustration downstream. Internal and external consistency can be checked through reviews, viewpoint resolution, various formal methods, and prototyping.

Ranking

A requirements set is ranked if the items are prioritized for importance and/or stability. Importance is a relative term, and its meaning needs to be resolved on a case-by-case basis. Stability means the likelihood that the requirement would change. For example, a hospital information system will always have doctors, nurses, and patients (but governing legislation will change). The ranking could use a numerical scale (positive integers or real numbers), a simple rating system (e.g., Mandatory, Desirable, Optional), or could be ranked by mission criticality.

For example, NASA uses a four-level ranking system. Level 1 requirements are mission-level requirements that are very high level and very rarely change. Level 2 requirements are "high level" with minimal change. Level 3 requirements are those requirements that can be derived from level 2 requirements. That is, each level 2 requirement traces to one or more level 3 requirement. Contracts usually bid at this level of detail. Finally, level 4 are detailed requirements and are typically used to design and code the system (Rosenberg et al.).

Ranking is an extremely important quality of an SRS document. Suppose in the course of system design two requirements cannot be met simultaneously. It becomes easier to decide which requirement to relax based on its ranking. In addition to being useful for tradeoff engineering, ranking can be used for cost estimation and negotiation, and for dispute resolution. Ranking validation is easy enough through reviews and viewpoint resolution (to agree upon the rankings).

Verifiability

An SRS is verifiable if satisfaction of each requirement can be established using measurement or some other unambiguous means. This quality is important because a requirement that cannot be shown to be met has not been met. When requirements cannot be measured, they cannot be met and disputes will follow.

Verifiability can be explored through various reviews, through test case design (design driven development), and through viewpoint resolution.

Modifiability

Modifiability means that the SRS and structure of the document will readily yield to changes. Usually this means that the document is numbered, stored in a convenient electronic format, and compatible with common document processing and configuration tools.

It is obvious why modifiability is an important quality of an SRS document—requirements will change! Ease of modification will also reduce costs, assist in meeting schedules, and facilitate communications. Reviews and inspections are the most obvious way to assess a document's modifiability.

Traceability

An SRS is traceable if each requirement is clearly identifiable, and all linkages to other requirements (e.g., dependencies) are clearly marked. Traceability is an essential quality for effective communications about requirements, to facilitate easy modification, and even for legal considerations. For example, in the case of a dispute, it is helpful to show that responsible linking of related requirements was done.

In addition, each requirement should have a link to at least one other requirement. Traceability can be measured using network-like analyses. For example, we could count the efferent (inward) and afferent (outward) coupling as indicated by the key phrases "uses," "is referenced by," "references," "is used by," and so on.

Generally, we would like each requirement to be tested by more than one test case. At the same time, we would like each test case to exercise more than one requirement. The "test span" metrics are used to characterize the test plan and identify insufficient or excessive testing:

- Requirements per test
- Tests per requirement

Research is still ongoing to determine appropriate statistics for these metrics. But at the very least, you can use these metrics to look for inconsistencies and non-uniform test coverage. Of course, there is always a tradeoff between time and cost of testing versus the comprehensiveness of testing. But testing is not the subject of this book.

Group reviews and inspections and automated tools can also be used to check for traceability between requirements to/from tests.

NASA Requirements Testing

One would think that the American space agency NASA is a place where rigorous requirements engineering is conducted. This is a correct assumption. Given that NASA is engaged in the engineering of very-high-profile, high-cost, and most importantly, life-critical systems, the techniques used and developed here are state of the art. Table 5.2 contains an excerpt from NASA Procedural Requirements for requirements engineering.

NASA is heavily invested in the use of formal methods for requirements verification and validation, and a number of techniques and tools for this purpose have been developed.

Notice that verification matrices are specifically mentioned as helping to accomplish the software requirements engineering goals. Requirements management, which will be discussed in Chapter 9, is specifically mentioned in the directive.

NASA ARM Tool

The NASA ARM Tool was developed at NASA's Software Assurance Technology Center at Goddard Space Flight Center in Greenbelt, MD. This tool conducts an analysis of the text of the SRS document and reports certain metrics. The metrics are divided into two categories: micro- and macro-level metrics. Micro-level indicators count the occurrences of specific keyword types. Macro-level indicators are coarse-grained metrics of the SRS documentation.

Micro-level indicators include

- imperatives
- continuances
- directives
- options
- weak phrases

Macro-level indicators include

- size of requirements
- text structure
- specification depth
- readability

These micro- and macro-level indicators will be described in some detail.

Table 5.2 Excerpt from NASA Procedural Requirements for Requirements Engineering (NASA)

3.1.1 Requirements Development
3.1.1.1 The project shall document the software requirements. [SWE-049]
Note: The requirements for the content of each Software Requirement Specification and Data Dictionary are defined in Chapter 5.
3.1.1.2 The project shall identify, develop, document, approve, and maintain software requirements based on analysis of customer and other stakeholder requirements and the operational concepts. [SWE-050]
3.1.1.3 The project shall perform software requirements analysis based on flowed-down and derived requirements from the top-level systems engineering requirements and the hardware specifications and design. [SWE-051]
Note: This analysis is for safety criticality, correctness, consistency, clarity, completeness, traceability, feasibility, verifiability, and maintainability. This includes the allocation of functional and performance requirements to functions and subfunctions.
3.1.1.4 The project shall perform, document, and maintain bi- directional traceability between the software requirement and the higher level requirement. [SWE-052]
Note: The project should identify any orphaned or widowed requirements (no parent or no child) associated with reused software.
3.1.2 Requirements Management
3.1.2.1 The project shall collect and manage changes to the software requirements. [SWE-053]
Note: The project should analyze and document changes to requirements for cost, technical, and schedule impacts.
3.1.2.2 The project shall identify inconsistencies between requirements, project plans, and software products and initiate corrective actions. [SWE-054]
Note: A verification matrix supports the accomplishment of this requirement.

In addition, various ratios can be formed using macro- and micro-level indicators. No particular thresholds for the metrics are given (research is still being conducted in this regard). However, at the end of this section, summary metrics for 56 NASA projects are given for comparison.

A description of the metrics and some excerpts from the ARM report for creating the Smart Home SRS document found in the appendix. The definitions are derived from those reported by the tool and described by the authors of the tool in a related report (Rosenberg et al.).

Imperatives

The first metric, imperatives, is a micro indicator that counts the words and phrases that command that something must be provided. Imperatives include

- "shall"—dictates the provision of a functional capability
- "must" or "must not"—establish performance requirements or constraints
- "will"—indicates that something will be provided from outside the capability being specified. For example, *"the building's electrical system will power the xyz system."*
- "is required to"—used in specifications statements written in passive voice
- "are applicable"—used to include, by reference, standards or other documentation as an addition to the requirements being specified
- "responsible for"—used as an imperative for systems whose architectures are already defined. For example, *"the xyz function of the abc subsystem is responsible for responding to the pdq inputs."*

A more precise specification will have a high number of "shall" or "must" imperatives relative to other imperative types. Note that the word "should" is not recommended for use in an SRS. From both a logical and legal point of view "should" places too much discretion in the hands of system designers.

The counts of imperatives found in the Smart Home SRS document are shown in Table 5.3. An excerpt of the imperatives capture from the ARM output is shown in Figure 5.2.

Table 5.3 ARM Counts of Imperatives Found in the Smart Home SRS Document

Imperatives	*Occurrences*
shall	308
must	0
is required to	0
are applicable	0
are to	1
responsible for	0
will	51
should	7
Total	367

Continuances

Continuances are phrases that follow an imperative and precede the definition of lower-level requirement specifications. Continuances indicate that requirements have been organized and structured. Examples of and counts of continuances found in the Smart Home SRS document are shown in Table 5.4. The symbol ":" is treated as a continuance when it follows an imperative and precedes a requirement definition.

These characteristics contribute to the ease with which the requirement specification document can be changed. Too many continuances, however, indicate

shall # 1: In Line No. 169, ParNo. 3.1.1., @ Depth 3
3.1.1 System SHALL operate on a system capable of multiprocessing.

will # 51: In Line No. 624, ParNo. 9.12.3., @ Depth 3
9.12.3 System SHALL allow users to record greeting message that WILL be played after user defined number of rings.

should # 7: In Line No. 557, ParNo. 9.8., @ Depth 2
In the future this SHOULD be extended such that any commands can be programmed to control any device or system interface by the SH.

be able to # 1: In Line No. 324, ParNo. 7.3., @ Depth 2
Occupants and users of the SH's system should Be ABLE TO monitor the home from anywhere they wish.

normal # 1: In Line No. 490, ParNo. 9.3.14., @ Depth 3
9.3.14 Hot tub cover shall close with button press or if no activity / motion is detected for some time range, and water displacement levels are NORMAL (no one in the tub).

provide for # 1: In Line No. 104, ParNo. 2., @ Depth 2
The summation and harmonization of all the six categories of the SH will PROVIDE FOR a truly rewarding living experience for the occupants of the SH.

easy to # 1: In Line No. 193, ParNo. 4.1.2., @ Depth 3
4.1.2 System shall be EASY TO use.

can # 1: In Line No. 124, ParNo. 2., @ Depth 2
Ensuring the existing structure CAN support the improvements

may # 1: In Line No. 180, ParNo. 3.1.9., @ Depth 3
3.1.9 System MAY contain separate SAN device for storage flexibility

Figure 5.2 Excerpt of ARM output for Smart Home requirements specification document.

Table 5.4 ARM Counts for Continuances in the Smart Home SRS Document

Continuance	*Occurrence*
below	0
as follows	0
following	0
listed	0
in particular	0
support	0
and	85
:	2
Total	87

Table 5.5 Directives Found in the Smart Home SRS

Directive	*Occurrence*
e.g.	0
i.e.	14
For example	0
Figure	0
Table	0
Note	0
Total	14

Table 5.6 Options and Their Counts Found in the Smart Home SRS Document

Option Phrases	*Occurrence*
can	7
may	23
optionally	0
Total	30

multiple, complex requirements that may not be adequately reflected in resource and schedule estimates.

Directives

The micro-indicator "directives" count those words or phrases that indicate that the document contains examples or other illustrative information. Directives point to information that makes the specified requirements more understandable. Typical directives and their counts found in the Smart Home SRS document are shown in Table 5.5. Generally, the higher the number of total directives, the more precisely the requirements are defined.

Options

Options are those words that give the developer latitude in satisfying the specifications. At the same time, options give less control to the customer. Options and their counts found in the Smart Home SRS document are shown in Table 5.6.

Weak Phrases

Weak phrases are clauses that are subject to multiple interpretations and uncertainty and therefore can lead to requirements errors. Use of phrases such as "adequate" and "as appropriate" indicate that what is required is either defined elsewhere or worse, the requirement is open to subjective interpretation. Phrases such as "but not limited to" and "as a minimum" provide the basis for expanding requirements that have been identified or adding future requirements. The counts of weak phrases for the Smart Home SRS document are shown in Table 5.7.

The total number of weak phrases is an important metric that indicates the extent to which the specification is ambiguous and incomplete.

Table 5.7 Weak Phrases for the Smart Home SRS Document

Weak Phrase	*Occurrence*
adequate	0
as appropriate	0
be able to	3
be capable of	0
capability of	0
capability to	0
effective	0
as required	0
normal	1
provide for	1
timely	0
easy to	1
Total	6

Table 5.8 Incomplete Words and Phrases Found in the Smart Home SRS Document

Incomplete Term	*Occurrence*
TBD	0
TBS	0
TBE	0
TBC	0
TBR	0
not defined	0
not determined	0
but not limited to	0
as a minimum	0
Total	0

Incomplete

The "incomplete" micro-indicator counts words that imply that something is missing in the document, for whatever reason (for example, future expansion, undetermined requirements). The most common incomplete notation is "TBD" for "to be determined."

Variations of "TBD" include

- TBD—"to be determined"
- TBS—"to be scheduled"
- TBE—"to be established" or "yet to be estimated"
- TBC—"to be computed"
- TBR—"to be resolved"
- "Not defined" and "not determined" explicitly state that a specification statement is incomplete.
- "But not limited to" and "as a minimum" are phrases that permit modifications or additions to the specification.

Incomplete words and phrases found in the Smart Home SRS document are shown in Table 5.8.

Leaving incompleteness in the SRS document is an invitation to disaster later in the project. While it is likely that there may be a few incomplete terms in a well-written SRS due to pending requirements, the number of such words should be kept to an absolute minimum.

Subjects

Subjects are a count of unique combinations of words immediately preceding imperatives in the source file. This count is an indication of the scope of subjects addressed by the specification.

The ARM tool counted a total of 372 subjects in the Smart Home SRS document.

Specification Depth

Specification depth counts the number of imperatives at each level of the document and reflects the structure of the requirements. The topological structure of requirements was discussed in Chapter 4. The numbering and specification structural counts for the Smart Home SRS document as computed by the NASA ARM tool are provided in Table 5.9.

These counts indicate that the SRS requirements hierarchy has a "diamond" shape in the manner of Figure 4.8c.

Table 5.9 Numbering and Specification Structure Statistics for Smart Home SRS Document in the Appendix

Numbering Structure		*Specification Structure*	
Depth	*Occurrence*	*Depth*	*Occurrence*
1	19	1	0
2	71	2	50
3	265	3	258
4	65	4	64
5	0	5	0
6	0	6	0
7	0	7	0
8	0	8	0
9	0	9	0
Total	420	**Total**	372

Readability Statistics

In Chapter 4 we discussed the importance of clarity in the SRS document. There are various ways to evaluate reading levels, but most techniques use some formulation of characters or syllables per words and words per sentence. For example, the Flesch Reading Ease index is based on the average number of syllables per word and of words per sentence. Standard writing tends to fall in the 60–70 range but apparently, a higher score increases readability.

The Flesch-Kincaid Grade Level index is supposed to reflect a grade-school writing level, so a score of 12, means that someone with a 12th grade education would understand the writing. But standard writing averages 7th to 8th grade, and a much higher score is not necessarily good—higher level writing would be harder to understand. The Flesch-Kincaid Grade level indicator is also based on the average number of syllables per word and on words per sentence. There are other grade level indicators as well (Wilson et al.).

The NASA ARM tool does not provide the ability to count these metrics, but most versions of the popular Microsoft Word can provide at least some relevant statistics. For example, the version of Word 2007 used to prepare this manuscript will compute various word, character, paragraph, and sentence counts and averages. It will also compute the Flesch Reading Ease and Flesch-Kincaid Grade level metrics (consult the user's manual or on-line help feature to determine how to compute such metrics for your word processor, if available). In any case, we used Word to calculate the statistics for the Smart Home SRS document and obtained the output shown in Figure 5.3.

The SRS document is assessed to be at a 12th grade reading level. The low number of sentences per paragraph (1.2) is an artifact of the way the tool counted each numbered requirement as a new paragraph.

Summary of NASA Metrics

To get some sense of proportion and relevance to the ARM indicators, Rosenberg and colleagues studied 56 NASA software systems ranging in size from 143 to 4772 lines of code (Rosenberg et al.). They used their tool to collect statistics about these systems, which are summarized in Table 5.10.

From Table 5.10 we notice that one specification was only 143 lines of text and that the longest was 28,000 lines of text. It is interesting to note from Table 5.10 that even NASA specifications have "TBDs" and often, many options. Validatable and verifiable are two additional properties in the table (Wilson et al.).

Aside from the obvious interpretations, why are these ARM indicators important? Table 5.11 gives us the intriguing answer—because there is a correlation between these indicators and the IEEE 830 qualities that we have already discussed. A cross reference of NASA indicators to IEEE 830 qualities is shown in Table 5.11.

Looking at the table it seems that the text structure and depth are indicators of internal consistency (but not external). The number of directives and weak phrases are correlated (positively and negatively, respectively) with correctness. All of the

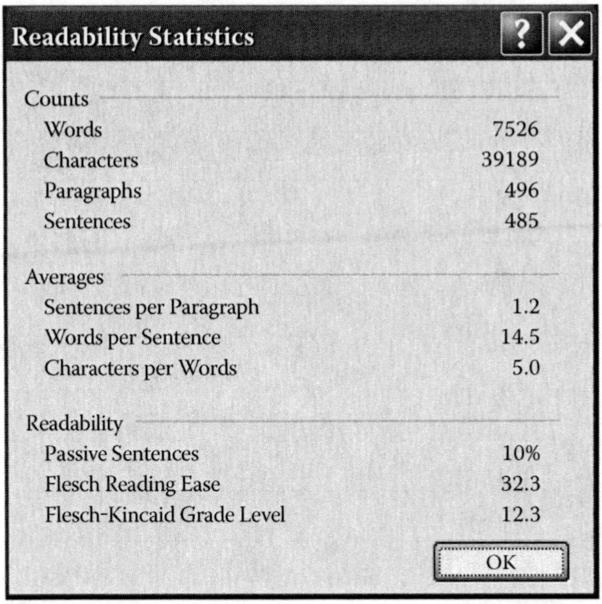

Readability Statistics	
Counts	
Words	7526
Characters	39189
Paragraphs	496
Sentences	485
Averages	
Sentences per Paragraph	1.2
Words per Sentence	14.5
Characters per Words	5.0
Readability	
Passive Sentences	10%
Flesch Reading Ease	32.3
Flesch-Kincaid Grade Level	12.3
	OK

Figure 5.3 Readability statistics for Smart Home SRS document obtained from Microsoft Office Word 2007.

Table 5.10 Sample Statistics from 56 NASA Requirements Specifications (Rosenberg)

	Lines of Text	*Imperatives*	*Continuances*	*Directives*	*Weak Phrases*	*TBD, TBS, TBR*	*Option (can, may…)*
Minimum	143	25	15	0	0	0	0
Median	2265	382	183	21	37	7	27
Average	4772	682	423	49	70	25	63
Max	28459	3896	118	224	4	32	130
Std Dev	759	156	99	12	21	20	39
Level 3 Specs	1011	588	577	10	242	1	5
Level 4 Specs	1432	917	289	9	393	2	2

Table 5.11 Cross Reference of NASA Indicators to IEEE 830 Qualities (Wilson)

	Indicators of Quality Attributes										
		Quality Attributes									
Categories of Quality Indicators	*Complete*	*Consistent*	*Correct*	*Modifiable*	*Ranked*	*Testable*	*Traceable*	*Unambiguous*	*Understandable*	*Validatable*	*Verifiable*
Imperatives	■			■			■	■	■	■	■
Continuances	■			■	■	■	■	■	■	■	■
Directives	■		■			■		■	■	■	■
Options	■					■		■	■	■	
Weak phrases	■		■			■		■	■	■	■
Size	■					■		■	■	■	■
Text structure	■	■		■	■		■		■		■
Specification depth	■	■		■			■		■		■
Readability				■		■	■	■	■	■	■

micro-indicators are linked to testability (both in a positive and negative correlation—you need "just enough" directives but not too many). Finally, all quality indicators (except for readability) contribute to completeness.

Exercises

5.1 What can be some pitfalls to watch out for in ranking requirements?

5.2 Explain how the following can help remove ambiguity from the SRS.
- formal reviews
- viewpoint resolution
- formal modeling

5.3 Which of the IEEE Standard 830 qualities seem most important? Can you rank these?

5.4 For an available SRS document, conduct an informal assessment of its IEEE 830 qualities.

5.5 For each Quality Attribute in Table 5.11 discuss its relationship to the Categories of Quality Indicators.

5.6 Should implementation risk be discussed with customers?
5.7 What are the advantages and risks of having requirements engineering conducted (or assisted) by an outside firm or consultants?
5.8 Create a traceability matrix for the SRS of appendix.
5.9 Calculate the Requirements per Test and Tests per Requirements metrics for the data shown in Table 5.1. Do you see any inconsistencies?
5.10 Consider the requirements in the SRS of the appendix:
   a. Which of these could be improved through the use of visual formalisms such as various UML diagrams?
   b. Select three of these and create the visual formalism.
5.11 Why is the NASA ARM tool useful in addition to objective techniques of SRS risk mitigation?
5.12 Install and run the NASA ARM tool on any available SRS document. What can you infer from the results?

References

Basili, V., G. Caldiera, and H.D. Rombach (1994) Goal question metric approach, *Encyclopedia of Software Engineering*, pp. 528–532, John Wiley & Sons, Inc.,

Boehm, B.W. (1984) Verifying and validating software requirements and design specifications, *IEEE Software*, (1): 75–88.

Hetzel, B. (1988) *The Complete Guide to Software Testing*, 2nd Edn, QED Information Sciences, Inc.

IEEE Std 830-1993, IEEE Recommended Practice for Software Requirements Specifications, Institute for Electrical and Electronics Engineers, Piscataway, NJ, 1993.

IEEE Std 1012-2004, IEEE Standard for Software Verification and Validation, Institute for Electrical and Electronics Engineers, Piscataway, NJ, 2004.

IEEE Std 12207.0-1996, IEEE/EIA Standard—Industry Implementation of International Standard ISO/IEC 12207:1995 (ISO/IEC 12207) Standard for Information Technology—Software Life Cycle Processes, Institute for Electrical and Electronics Engineers, Piscataway, NJ, 1996.

Laplante, P.A., W.W. Agresti, and G.R. Djavanshir (2007) Guest editor's introduction, special section on IT Quality Enhancement and Process Improvement, *IT Professional*, November/December, pp. 10–11.

NASA Procedural Requirements, NASA Software Engineering Requirements, NPR 7150.2, http://nodis3.gsfc.nasa.gov/displayDir.cfm?Internal_ID=N_PR_7150_0002_&page_name=Chapter3, 27 September 2004, last accessed 1 February 2008.

O'Neil, R. (2004) *Pennsylvania's "No Jake Braking" Signs*, OLR Research Report #004-R-0515, July 1, 2004. Found on the Web at http://www.cga.ct.gov/2004/rpt/2004-R-0515.htm, last accessed 20 June 2008.

Rosenberg, L., T. Hammer, and L. Huffman, Requirements, Testing, & Metrics, NASA Software Assurance Technology Center Report, satc.gsfc.nasa.gov, last accessed 1 February 2008.

Voas, J., and P. Laplante End brake retarder prohibitions: Defining "shall not" requirements effectively, *Computer* 2009.

Wilson, W.M., L.H. Rosenberg, and L.E. Hyatt, Automated analysis of requirement specifications, http://satc.gsfc.nasa.gov/support/ICSE_MAY97/arm/ICSE97-arm.htm, last accessed 1 February 2008.

Chapter 6

Formal Methods

Motivation

Systems have tremendous sensitivity to errors in a requirements specification—even a misplaced comma can have severe consequences. In code implementation, it is obvious that misplacing even a single character can make a great deal of difference. In fact, the accidental substitution of a period for a comma in a single Fortran statement resulted in the loss of the Mariner 1, the first American probe to Venus in 1962. But how can such a serious problem exist when misplacing a character in a requirements specification? Let's see how that might be so.

The title of Lynne Truss's 2004 book on punctuation, *Eats Shoots and Leaves,** could refer to either

- a panda, if the punctuation is as published, or
- a criminal who refuses to pay his restaurant bill if a comma is added after the word "eats."

Clearly the title of the book is not that of a software specification, but we hope this anecdote illustrates that simple punctuation differences can convey a dramatically different message or intent, especially in a requirements specification.

* Actually, the author heard the joke differently from members of the Royal Australian Air Force almost 20 years ago. As told, the Panda is an amorous, but lazy creature and is well known for its ability to scope out a tree occupied by a panda of the opposite sex, where it "eats, roots, shoots, and leaves." In a double entendre, the middle portion of the quote refers to the act of procreation.

Aside from punctuation, there are a number of problems with conventional software specifications built using only natural language and informal diagrams. These problems include ambiguities, where unclear language or diagrams leave too much open to interpretation; vagueness, or insufficient detail; contradictions, that is, two or more requirements that cannot be simultaneously satisfied; incompleteness or any other kind of missing information; and mixed levels of abstraction where very detailed design elements appear alongside high-level system requirements. To illustrate, consider the following hypothetical requirement for a missile launching system.

> 5.1.1 If the LAUNCH-MISSILE signal is set to TRUE and the ABORT-MISSILE signal is set to TRUE then do not launch the missile, unless the ABORT-MISSILE signal is set to FALSE and the ABORT-MISSILE-OVERRIDE is also set to FALSE, in which case the missile is not to be launched.

Aside from the fact that this requirement is written in a confusing manner, the complexity of the logic makes it difficult to know just exactly what the user intends. And if user's intent is wrongly depicted in the language of the requirements, then the wrong system will be built. It is sometimes said "syntax is destiny."

It is because we need precision beyond that which can be offered by natural languages that we frequently need to reach for more powerful tools, which can only be offered by mathematics.

What Are Formal Methods?

Formal methods involve mathematical techniques. To be precise, we look at three definitions for "formal methods":

Definition 1: [Encyclopedia of Software Engineering, 1994]

> "A method is formal if it has a sound mathematical basis, typically given by a formal specification language. This basis provides a means of precisely defining notions like consistency and completeness, and, more relevant, specification, implementation, and correctness."

Definition 2: [IEEE Standard Glossary, 1990]

> "1. A specification written and approved in accordance with established standards; 2. A specification written in a formal notation often for use in proof of correctness."

Definition 3: [Dictionary of Computer Science, Engineering, and Technology, 2001]

"a software specification and production method based on a precise mathematical syntax and semantics that comprises:

- A collection of mathematical notations addressing the specification, design, and development phases of software production, which can serve as a framework or adjunct for human engineering and design skills and experience
- A well founded logical inference system in which formal verification proofs and proofs of other properties can be formulated
- A methodological framework within which software may be developed from the specification in a formally verifiable manner

Formal methods can be operational, denotational, or dual (hybrid)."

It is clear from the three different but similar definitions that formal methods have a rigorous, mathematical basis.

Formal methods differ from informal techniques, such as natural language, and informal diagrams like flowcharts. The latter cannot be completely transliterated into a rigorous mathematical notation. Of course, all software requirements specifications will have informal elements, and there is nothing wrong with this fact. However, there are apt to be elements of the system specification that will benefit from formalization.

Techniques that defy classification as either formal or informal because they have elements of both are considered to be semi-formal. For example, the UML version 1.0 is considered to be semi-formal because not all of its meta models have precise mathematical equivalents.

But what advantage is there in applying a layer of potentially complex mathematics to the already complicated problem of behavioral description? The answer is given by Thomas E. Forster, a giant of formal methods: "One of the great insights of twentieth-century logic was that, in order to understand how formulae can bear the meanings they bear, we must first strip them of all those meanings so we can see the symbols as themselves … [T]hen we can ascribe meanings to them in a systematic way … That makes it possible to prove theorems about what sort of meanings can be borne by languages built out of those symbols." (Forster 2003).

A Little History

Formal methods have been in use by software engineers for quite some time. The Backus-Naur (or "Normal") Form (BNF) is a mathematical specification language originally used to describe the Algol programming language in 1959. Since then a number of formal methods have evolved. These include

- The Vienna Development Method (VDM), 1971
- Communicating Sequential Process (CSP), 1978
- Z (pronounced "zed"), late 1970s
- Pi-calculus, early 1990s
- B, late 1990s
- and many others

Finite State Machines, Petri Nets, and Statecharts or other techniques regularly used by systems and software engineers can be used formally. Other formal methods derive from general mathematical frameworks, such as category theory, and a number of domain-specific and general languages have been developed over the years, often with specialized compilers or other toolsets.

Formal methods are used throughout Europe, particularly in the UK, but not as widely in the United States. However, important adopters of formal methods include NASA, IBM, Lockheed, HP, and AT&T.

Using Formal Methods

Formal methods are used primarily for systems specification and verification. Users of UML 2.0 could rightly be said to be employing formal methods, but only in the specification sense. The languages B, VDM, Z, Larch, CSP, and Statecharts, and various temporal logics are typically used for systems specification. And while the clarity and precision of formal systems specification are strong endorsements for these, it is through verification methods that formal methods really show their power. Formal methods can be used for two kinds of verification: theorem proving (for program proving) and model checking (for requirements validation). We will focus on the former, however.

Formal methods can be used in any setting, but they are generally used for safety critical systems, for COTS validation/verification (e.g., by contract), for high financial risk systems (e.g., in banking and securities), and anywhere that high-quality software systems are needed.

Formal methods activities include writing a specification using a formal notation, validating the specification, and then inspecting it with domain experts. Further, one can perform automated analysis using theorem (program) proving or refine the specification to an implementation that provides semantics-preserving transformations to code. Finally, you can verify that the implementation matches the specification (testing).

Formal Methods Types

There are several formal methods types. The first, model-based, provide an explicit definition of state and operations that transform the state. Typical model-based formal methods include Z, B, and the Vienna Development Method. The next, algebraic methods, provide an implicit definition of operations without defining state. Algebraic methods include Larch, PLUSS, and OBJ. Process algebras provide explicit models of concurrent processes—representing behavior with constraints on allowable communication between processes. Among these are CSP and CCS. Logic-based formal methods use logic to describe properties of systems. Temporal and interval logics fall into this category. Finally, net-based formal methods offer implicit concurrent models in terms of data flow through a network, including conditions under which data can flow from one node to another. Petri-nets are a widely used net-based formal method.

It should be noted that formal methods are different from mathematically based specifications. That is, specifications for many types of systems contain some formality in the mathematical expression of the underlying algorithms. Typical systems include process control, avionics, medical, and so on. Such use of mathematics does not constitute use of formal methods, though such situations may lend themselves to formal methods.

Examples

To illustrate the use of formal methods in requirements engineering, we present a number of examples using several different techniques. Our purpose is not to present any one formal method and expect the reader to master it. Rather, we wish to show a sampling of how various formal methods can be used to strengthen requirements engineering practice.

Formalization of Train Station in B

In their most straightforward use, formal methods can express system intent. In this use of formal methods, we exploit the conciseness and precision of mathematics to avoid the shortcomings of natural languages and informal diagrams. B is a model-based formal language that is considered an "executable specification" (collection of abstract machines) that can be translated to C++ or Ada source code. In this case, we use the modeling language B to provide a specification for a train station, similar to Paris's Line 14 (Meteor) and New York City's L Line (Canarsie). The model used is based on an example found in Lano (1996). Those familiar with Z will see a great deal of similarity in the structure and notation to B. Thanks to Dr. George Hacken of the New York Metropolitan Transportation Authority for providing this example, which is derived from a 1999 talk he gave along with Sofia Georgiadis to the New Jersey Section of the IEEE Computer Society.

The model begins as shown in Figure 6.1. The specification starts with the name of the machine and any included machines (in this case, **TrackSection**, which is not shown here). Then a list of variables is given and a set of program

```
TRAIN STATION
MACHINE Station
INCLUDES TrackSection

VARIABLES
    platforms, max_trains, trains_in_station

INVARIANT
    trains_in_station ∈ N ^
    max_trains ∈ N ^ trains_in_station ≤ max_trains ^
    platforms ∈ seq(sections) ^
    size(platforms) = card(ran(platforms)) ^
    max_trains ≤ size(platforms)
```

Figure 6.1 Setup specification for train station.

INVARIANTs—conditions that must hold throughout the execution of the program. Here **N** represents the natural numbers {0,1,2,3, ...}. We continue with a partial description of the behavioral intent of this model.

Notice that the three variables of interest are the **platforms, max_trains**, and **trains_in_station**. The **INVARIANT** sets constraints on the number of trains in the station (must be a non-negative number) and the maximum number of trains in the system (must be a non-negative number and there can be no more than that number in the station). It continues with platforms as a sequence of train sections, in the range of the number of sections, which acts as an upper bound on the maximum number of trains (since only one train can occupy each section of track).

The next part of the B specification is the initialization section (Figure 6.2).

This specification initializes the platform sequence to null and the maximum number of trains and trains in a station to zero.

Now we introduce a set of operations on the train station dealing with arrivals, departures, openings and closings, and resetting the maximum number of trains. For example, the arrival specification is shown in Figure 6.3. A precondition (**PRE**) before train **ts** can arrive is that there is room in the station for the train and that a platform is available. If so, mark the train as arriving and increment the count of trains in the station.

The next operation involves train departure (Figure 6.4). The precondition is that the train is in the range of the sequence of platforms (on a platform) and that the train is not moving (blocked). If so, then train **ts** is marked as having departed, and the number of trains in that station is decreased by one.

```
INITIALIZATION

    platforms, max_trains, trains_in_station := [], 0, 0
```

Figure 6.2 Initialization specification for train station.

```
OPERATIONS

  train_arrives(ts)≅
    PRE trains_in_station < max_trains ^ ts∈ ran(platforms) ^

        tstate(ts) = free

    THEN

        arrival(ts) | |

        trains_in_station := trains_in_station + 1

    END;
```

Figure 6.3 Train arrival behavior for station.

```
train_departs(ts)≅

    PRE ts ran(platforms) ^

        tstate(ts) = blocked

    THEN

        departure(ts) | |

        trains_in_station := trains_in_station - 1

    END;
```

Figure 6.4 Train departure behavior for station.

The next operation is the opening platform **ts** (Figure 6.5). The precondition is that the platform is in the range of the sequence of platforms for that station and that the current platform is closed. If so, then platform **ts** is marked as being opened.

The closing of a platform operation is similar (Figure 6.6). The precondition is that the platform is in the range of the sequence of platforms for that station and that the current platform is open. If so, then platform **ts** is marked as being closed.

Finally, an operation is needed to set the maximum number of trains, **mt**, that can be at a given station (Figure 6.7). The preconditions are that **mt** must be a natural number and cannot exceed the maximum number of platforms and that the maximum number of trains for a station cannot be less than the number of trains already there.

The point behind this is that the mathematical specification is far more precise than the natural language "translation" of it. In addition, the mathematical representation of the behavior is less verbose than the natural language version. Finally, the

```
open_platform(ts)≅
        PRE ts ∈ran(platforms)
            tstate(ts) = closed
THEN
            open(ts)
        END;
```

Figure 6.5 Platform opening behavior for station.

```
close_platform(ts)≅
    PRE ts ∈ran(platforms)
        tstate(ts) = free
    THEN
        close(ts)
    END;
```

Figure 6.6 Platform closing behavior for station.

```
set_max_trains(mt)≅
    PRE mt∈ N ^ mt ≤ size(platforms) ^
        trains_in _station ≤ mt
    THEN
        max_trains := mt
    END
```

Figure 6.7 Behavior for setting the maximum number of trains for a train station.

representation in B enables correctness-proofs, via standard predicate calculus. And as stated before, the B specification can actually be converted to C++ or Ada code.

Formalization of Space Shuttle Flight Software Using MurΦ

The next example of formalization uses the Prototype Verification System (PVS) language to model and prove various aspects of the space shuttle flight software system (Crow and Di Vita 1998). PVS is a state-driven language, meaning it implements a finite state machine to drive behavior.

A finite state machine model, M, consists of set of allowable states, S, a set of inputs, I, set of outputs, O, and a state transition function, described by Equation 6.1.

$$M : I \times S \rightarrow [O \times S] \tag{6.1}$$

The initial and terminal states need to be defined as well.

Part of the implementation code in PVS is shown in Figure 6.8. Without explicating the code, it is interesting to point out its expressiveness in modeling the finite state machine.

Here the **principal_function** defines an interface incorporating the inputs (**pf_inputs**) and states (**pf_state**) needed to drive the behavior. The output is an output expression and the next state in the machine.

The behavior specification continues as an infinite sequence of state transitions (functional transformations) (Figure 6.9, top part) and as a set of functions associated with each state—in this case, on the state associated with firing small control jet rockets (vernier rockets). Even without fully understanding the syntax of PVS, because it is similar to C, C++, or Java, we hope you can see how this behavior of the system can be clearly described.

In addition to clarity and economy, another advantage of using PVS to describe the behavior is that an interactive proof checker and Stanford University's MurΦ (pronounced "Murphy") can be used for theorem proving (specification verification).

```
pf_result: TYPE = [# output: pf_outputs, state:  pf_state #]

principal_function (pf_inputs, pf_state,
                    pf_I_loads, pf_K_loads,
                    pf_constants) : pf_result =

    (# output := <output expression>,
       state   := <next-state expression> #)
```

Figure 6.8 Initialization code in PVS for space shuttle functionality (Crow and Di Vita 1998).

```
finite_sequences[n: posnat, t: TYPE] : THEORY
BEGIN

   finite_seq: TYPE = [below[n] -> t]

END finite_sequences

requirements: THEORY
BEGIN
IMPORTING finite_sequences
...

  jet: TYPE
  rot_jet: TYPE FROM jet

  jet_bound: posint
  jet_count: [rot_jet -> below[jet_bound)]
  jet_count_ax: AXIOM injective?(jet_count)

  finite_number_of_jets: LEMMA . . .

  vernier?: pred[rot_jet]
  vernier_jet: TYPE = (vernier?)
  there_is_a_vernier: AXIOM (EXISTS j: vernier? (j)

  primary_rot? (j): bool = NOT vernier? (j)
  primary_rot_jet: TYPE = (primary_rot?)
  there_is_a_primary: AXIOM
     (EXISTS j: primary_rot? (j))

  downfiring?: pred[vernier_jet]

...
END requirements
```

Figure 6.9 Behavior of vernier rockets in space shuttle system (Crow and Di Vita 1998).

*Formalization of an Energy Management System Using Category Theory**

Category Theory (CT) permits functional notation to be used to model abstract objects and is therefore useful in formal specification of those objects (Awodey 2005). This formalization facilitates good design and can assist in user interface design and permits users of the system to build on their domain knowledge as they learn the constraints that affect design decisions. These design decisions then become evident as the rules are applied from requirements to design to user interface

* This section is adapted from Formalization of an EMS System Using Category Theory, Master's thesis, Ann Richards under the supervision of P. Laplante (Penn State University, 2006).

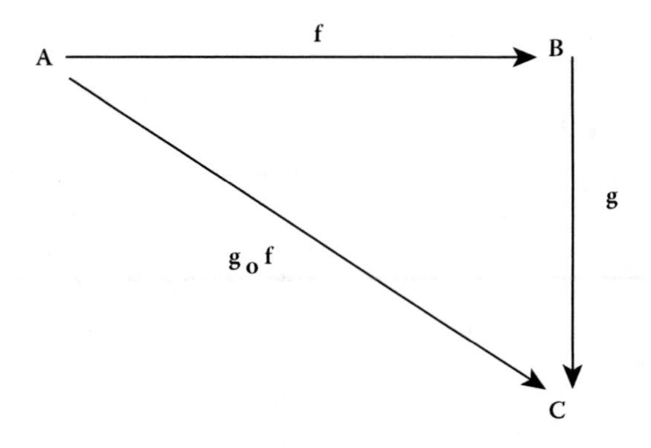

Figure 6.10 A simple functional relationship.

functionality. Finally, for ease of maintenance and troubleshooting for the designers and programmers, the use of CT can show where an error might be contained, while allowing for the group of functions to be analyzed and troubleshooting on an input-by-input basis. The following is a brief introduction to category theory for software specification.

Suppose we have classes of abstract objects, A, B, and C. Further, suppose there are functions f and g such that and $f: A \rightarrow B$ and $g: B \rightarrow C$. A category is composed of the objects of the category and the morphisms of the category. A basic functional relationship can be seen in Figure 6.10 where $f \circ g$ is the composition of f and g.

CT is widely used to model structures and their transformations over time (represented by finite or infinite sequences of functional compositions). For example, in Figure 6.11, the sequence of composed relationships between f and g would be given by gf, gfgf, gfgfgfgf, and so on.

These sequences also represent categories. For example, we just saw the space shuttle behavioral specification represented as an infinite sequence of state transitions using the PVS notation.

The composition of a system is also appropriate to model the modularity of a larger system and its interconnections. In comparison to a truth table, which is a great formalizing and testing tool, CT can associate related items and then decompose each into something that can be examined on a lower level. For example, business logic can be placed in a category either intuitively or architecturally.

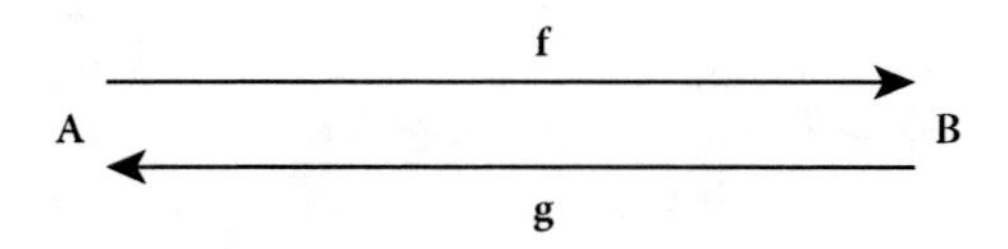

Figure 6.11 An infinite category.

Example: An Energy Management System

The example we describe is a partial formalization of a power company's energy management system. Any power generation utility has to deal with a complex, real-time system involving a high level of automation, fault tolerance, and redundancy. Security vulnerability is also of great concern, a fact that was recently highlighted—certain individuals hacked into computer systems of utility companies outside the US, causing a power outage that affected multiple cities. The attacks were based on vulnerabilities in SCADA (supervisory control and data acquisition) systems, on which many energy management systems run (including the one described here). It is possible that a thorough analysis of the system using formal methods could have identified the vulnerabilities.

In an energy management system (EMS) an open access gateway (OAG) serves as a communication liaison for many communication protocols between computer systems and remote terminal units (RTU). The OAG is also used to talking to the plant digital interface (PDI) via intercompany communications protocol (ICCP) and to other entities such as the interconnection and the TMS as shown in Figure 6.12.

To formalize Figure 6.14 using category theory, we first translate the entities using shorthand symbols shown in Figure 6.13.

The resultant relabeled system is shown in Figure 6.14. Functional notations have been added to Figure 6.14, but these would need to be defined further.

We begin with EMS A and EMS B from Figure 6.14. If A1 = EMS A and B1 = EMS B, then we have $f1 : A1 \rightarrow B1$. Next, $f1$ is defined on all $A1$ and all the values of $f1 \in B1$. That is, $range(f1) \subseteq B1$. In other words $f1$ is a bi-directional mapping from EMS A and EMS B and this function clearly describes communications between CFGCTRL and CFGMONI.

Continuing, there is another function $g1 : A1 \rightarrow D1$ and an associated composed function $g1 \circ f1 : A1 \rightarrow C1 \cap D1$, that is, $(g1 \circ f1)(a1) = g1(f1(a1))$ and $a1 \in A1$. The composed function is associative, as follows: If there is another function $j : C1 \rightarrow A2$ and forming $j \circ g1$ and $g1 \circ f1$ then we can compare $j \circ g1 \circ f1$ and $j \circ (g1 \circ f1)$. As indicated in Figure 6.14 it becomes apparent that these two functions are always identical.

At this point, it should be noted that the set of all inputs and outputs from $A1 \cap B1 \subset C1 \cup D1$ where $C1$ and $D1$ is CfgMoni on both the EMS A and B servers. The set of inputs and outputs from $C1 \cap D1 \subset A1 \cup B1$ where $A1$ and $B1$ is the CfgCtrl on the EMS A and B servers. CfgCtrl communicates with itself on both the EMS and OAG servers. The function $j(g) : C1 \rightarrow A2 \cup B2$ shows the configuration from the EMS CfgMoni to OAG CfgCtrl. OAG also communicates with CfgCtrl on its home server, that is, there is a function $g1(g) : C1 \rightarrow A1$. This feature is needed because CfgCtrl communicates with each redundant server so that controls can be issued while CfgMoni sends and receives status to CfgCtrl for up-to-date status on the server that it is monitoring.

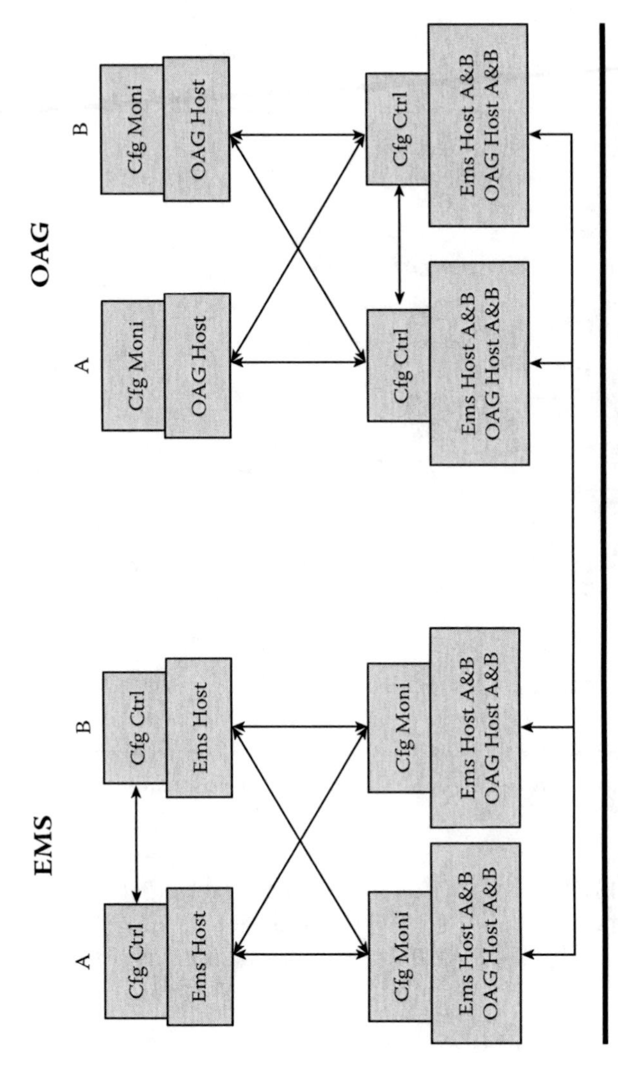

Figure 6.12 Configuration Monitor (CfgMoni) and Configuration Control (CfgCTRL) EMS and OAG relationship.

A1 – EMSA CFGCTRL	A2 – OAGA CFGCTRL
B1 – EMSB CFGCTRL	B2 – OAGB CFGCTRL
C1 – EMSA CFGMONI	C2 – OAGA CFGMONI
D1 – EMSB CFGMONI	D2 – OAGB CFGMONI
f1 – EMS CFGCTRL	f2 – OAG CFGCTRL
g1 – EMS CFMONI	g2 – OAGEMS CFMONI
j(g)– EMS CFGMONI to OAG CFGCTRL	h(g)–OAG CFGCTRL to EMS CFGMONI

Figure 6.13 Data dictionary and category reference.

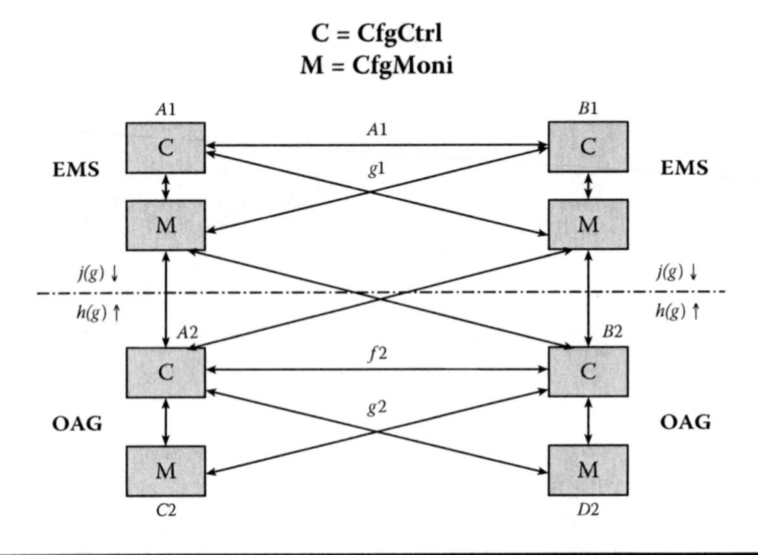

Figure 6.14 Partial formalization of Figure 6.12 using relabeled components.

To relate this to the EMS and OAG configurations, the set $A1$ is EMS CfgCtrl A, the set $B1$ is EMS CfgCtrl B, and $j(g) : (C1 \cup D1) \rightarrow (A2 \cup B2)$ shows that CfgMoni from either EMS A or B maps to CfgCtrl on the OAG A or B. And we could continue with this formalization (only a partial formalization has been presented).

Requirements Validation

We already discussed requirements validation in Chapter 5 using informal techniques, but formal techniques are particularly appropriate for testing the requirements to ensure, for example, that the requirements are consistent. By consistent we mean that the satisfaction of one requirement does not preclude the satisfaction of any other. One way to formally prove consistency is with a truth table. In this case, we rewrite each requirement as a Boolean proposition. Then we show that there is some combination of values for the Boolean variables of this collection of requirements for which all propositions (requirements) are true.

To illustrate, consider the following requirements from part of the pet store POS system:

1.1 If the system software is in debug mode, then the users are not permitted to access the database.
1.2 If the users have database access, then they can save new records.
1.3 If the users cannot save new records, then the system is not in debug mode.

We convert the requirements into a set of Boolean propositions. Let, p, q, and r be Boolean variables and let

p: be the statement "the system software is in debug mode"
q: be the statement "the users can access the database"
r: be the statement "the users can save new records"

Clearly, the specifications are equivalent to the following propositions:

1.1 $p \Rightarrow \neg q$
1.2 $q \Rightarrow r$
1.3 $\neg r \Rightarrow \neg p$

Now we construct a truth table (Table 6.1) and determine if any one of these rows has all "T"s in the columns corresponding to propositions 1.1, 1.2, and 1.3, meaning each of the requirements is satisfied. If we can find such a row, then the requirements are consistent.

We see that there are four rows where all three propositions representing the requirements are true for all combinations of Boolean values. Therefore, this set of requirements is consistent.

Notice that requirement 1.3 likely does not make logical sense. If the users cannot save new records then it is likely it was intended to say that the system is in debug mode (as opposed to not in debug mode). We deliberately left this logically questionable requirement in place to make a point—this system of requirements is consistent but not necessarily correct (in terms of what the customer intended). This consistency checking process will not uncover problems of intent (validity)—these must be uncovered through other means such as requirements reviews.

Consistency checking using Boolean satisfiability is a powerful tool. However, although the process can be automated, the problem space grows large very quickly. For n logical variables in the requirements set, the problem is $O(2^n)$. This kind of problem quickly becomes intractable even for supercomputers—it is a well-known NP-Complete problem, the Boolean Satisfiability problem. Therefore, we would

Table 6.1 Truth Table Proof of Consistency for a Collection of Requirements

p	Q	r	$\neg p$	$\neg q$	$\neg r$	$p \Rightarrow \neg q$	$q \Rightarrow r$	$\neg r \Rightarrow \neg p$
T	T	T	F	F	F	F	T	T
T	T	F	F	F	T	F	F	F
T	F	T	F	T	F	**T**	**T**	**T**
T	F	F	F	T	T	T	T	F
F	T	T	T	F	F	**T**	**T**	**T**
F	T	F	T	F	T	T	F	T
F	F	T	T	T	F	**T**	**T**	**T**
F	F	F	T	T	T	**T**	**T**	**T**

likely use this technique for requirements verification only for highly mission-critical situations. For example, the launch/no-launch decision logic for a weapon, dosage administration logic for some kind of medical equipment, shut-down logic for a nuclear power plant, and so on.

Theorem Proving

Theorem proving techniques can be used to demonstrate that specifications are correct. That is, axioms of system behavior can be used to derive a proof that a system (or program) will behave in a given way. Remember, a specification and program are both the same thing—a model of execution. Therefore program proving techniques can be used for appropriately formalized specifications. These techniques, however, require mathematical discipline.

Program Correctness

A system is correct if it produces the correct output for every possible input and if it terminates. Therefore, system verification consists of two steps:

- Show that, for every input, the correct output is produced (this is called partial correctness).
- Show that the system always terminates.

To deal with this we need to define assertions. An assertion is some relation that is true at that instant in the execution of the program. An assertion preceding a statement in a program is called a precondition of the statement. An assertion following a statement is called a postcondition. Certain programming languages, such as Eiffel and Java, incorporate run-time assertion checking. Assertions are also used for testing fault-tolerance (through fault-injection).

Hoare Logic

Much of the following follows the presentation found in Gries and Gries (2005). Suppose we view a requirements specification as a sequence of logical statements specifying system behavior. In 1969 C.A.R (Tony) Hoare introduced the notation (called the Hoare Triple):

$$P\ \{S\}\ Q$$

where P and Q are assertions (pre- and post-conditions, respectively) and S is a system behavioral segment. Then, the precondition, statement, postcondition triple has the following meaning: Execution* of the statement begun with the precondition true is guaranteed to terminate, and when it terminates, the postcondition will be true.

The intent is that P is true before the execution of S, then after S executes, Q is true. Note that Q is false if S does not terminate.

Hoare's logic system can be used to show that the system segment (specification) is correct under the given assertions. For example, suppose we have the following:

```
//{x = 5}        an assertion (pre-condition)

z = x + 2;

//{z = 7}        an assertion (post-condition)
```

Proof: Suppose that {x = 5} is true, then z = 5 + 2 = 7 after the system begins execution. So the system is partially correct. That the system terminates is trivial, as there are no further statements to be executed. Hence, correctness is proved.

Hoare added an inference rule for two rules in sequence. We write this as:

$$\frac{P\ \{S_1\}\ Q \qquad Q\ \{S_2\}\ R}{P\ \{S_1;\ S_2\}\ R}$$

* "Execution" means either program execution or effective change in system behavior in the case of a non-software behavioral segment.

This inference rule means that if the postcondition (Q) of the first system segment (S_1) is the antecedent (precondition) of the next segment, then the postcondition of the latter segment (R) is the consequent of the concatenated segments. The horizontal line means "therefore" or, literally, "as a result."

To illustrate, we show that the following specification is correct under the given assertions:

```
//{x = 1 ∧ y = 2}   an assertion (pre-condition)

x = x + 1;

z = x + y;

//{z = 4}           an assertion (post-condition)
```

Proof: Suppose that {x = 1} is true, then the system executes the first instruction, x = 2. Next, suppose that {y = 2} is true. Then after the execution of the second statement z = 2 + 2 = 4. Hence z = 4 after the execution of the last statement and the final assertion is true, so the system is partially correct. That the system terminates is trivial, as there are no further statements to be executed.

An inference rule is needed to handle conditionals. This rule shows that

$$(P \wedge \text{condition})\{S\}Q$$

$$(P \wedge \neg\, \text{condition}) \Rightarrow Q$$

$$\overline{\therefore\ P\ \{\text{if condition then } S\}Q}$$

To illustrate, we show that the specification segment is correct under the given assertions:

```
//{Any}        an assertion (pre-condition) that is always true

if x > y then

    y = x;

//{y ≤ x}      an assertion (post-condition)
```

Proof: If, upon entry to the segment, $y < x$, then the if statement fails and no instructions are executed and clearly, the final assertion must be true. On the other hand, if upon entry, $y > x$, then the statement $y = x$ is executed. Subsequently $y = x$ and the final assertion $y \le x$ must be true. That the system terminates is obvious, as there are no further statements to be executed.

Another inference rule handles if-then-else situations, that is,

$$(P \wedge \text{condition})\{S_1\}Q$$

$$(P \wedge \neg\ \text{condition})\{S_2\}Q$$

$$\overline{\qquad\qquad}$$

$$\therefore\ P\ \{\text{if condition then } S_1 \text{ else } S_2\}Q$$

To illustrate, we show that the specification segment is correct under the given assertions:

```
//{Any}          an assertion (pre-condition) that is always true

if x < 0 then

    abs = –x;

else

    abs = x;

//{abs = |x|}    an assertion (post-condition)
```

Proof: If $x < 0$, then $abs = -x \Rightarrow abs$ is assigned the positive x. If $x \ge 0$ then *abs* is also assigned the positive value of x. Therefore $abs = |x|$. That the system terminates is trivial, as there are no further statements to be executed.

Finally, we need a rule to handle "while" statements:

$$(P \wedge \text{condition})\{S\}P$$

$$\overline{\qquad\qquad}$$

$$\therefore\ (P \wedge \text{while condition})\{S\}(\neg\text{condition} \wedge P)$$

We will illustrate this rule through example shortly.

So far the specification snippets (one might say they are really "code' snippets) that have been proven represent very simplistic but low-level, detailed behavior unlikely to be found in a requirements specification. Or is this the case? It is not so hard to imagine that for critical behavior, it might be necessary to provide specifications at this level of detail. For example, the behavior required for the dosing logic for an insulin pump (or machine that delivers controlled radiation therapy) might require such logic. The decision to launch a missile, control the life support system in the Space Station, or shut down a nuclear plant might be based on simple, but critical logic. This logic, incidentally, could have been implemented entirely in hardware.

Even our running examples might have some critical logic that needs formal proof. For example, the baggage counting logic for the baggage handling system or certain inventory control logic might require the following behavior.

```
//{sum = 0 ∧ count = n ≥ 0}

while count > 0

    {

    sum = sum + count;

    count = count--;

    }

//{sum = n(n + 1)/2}
```

And so it would be necessary to show that the specification segment is correct under the given assertions. To prove it, we use induction.

Basis:

Suppose that sum = 0 and n = 0 as given by the assertion. Now upon testing of the loop guard, the value is false and the system terminates. At this point the value of sum is zero, which satisfies the postcondition. That the system terminates was just shown.

Induction hypothesis:

The program is correct for the value of count = n. That is, it produces a value of $sum = n(n + 1)/2$ and the system terminates.

Induction step:

Suppose that the value of sum = 0 and count = n + 1 in the precondition.

Upon entry into the loop, we assign the value of $sum = n + 1$, and then set $count = n$. Now, from the induction hypothesis, we know that for $count = n$, $sum =$

$n(n + 1)/2$. Therefore, upon continued execution of the system, sum will result in $(n + 1) + n(n + 1)/2$, which is just $(n + 1)(n + 1 + 1)/2$ and the system is partially correct.

Since by the induction hypothesis, by the nth iteration of the loop count has been decremented n times (since the loop exited when count was initialized to n). In the induction step, count was initialized to $n + 1$, so by the nth iteration it has a value or 1. So after the n + 1st iteration it will be decremented to 0, and hence the system exits from the loop and terminates.

Does this approach really prove correctness? It does, but if you don't believe it, try using Hoare logic to prove an incorrect specification segment to be correct, for example:

```
//{sum = 0 ∧ count = n ≥ 0}

while count < n

        {

        sum = sum + count;

        count = count++;

        }

//{sum = n(n + 1)/2}
```

This specification is incorrect because "n" is left out of the sum and it will defy proof (unless you cheat).

It is easy to show that a for loop uses a similar proof technique to the while loop. In fact, by Böhm and Jacopini (1966), we know we can construct verification proofs just from the first two inference rules. For recursive procedures, we use a similar, inductive, proof technique applying the induction to the n + 1st iteration in the induction step, and using strong induction if necessary.

Model Checking

Given a formal specification, a model checker can automatically verify that certain properties are theorems of the specification. Model checking has traditionally been used in checking hardware designs (e.g., through the use of logic diagrams), but it has also been used with software. Model checking's usefulness in checking software specifications has always been problematic due to the combinatorial state explosion and because variables can take on non-Boolean values. Nevertheless, Edmund M. Clarke, E. Allen Emerson, and Joseph Sifakis won the 2007 A. M. Turing Award for their body of work making model checking more practical, particularly in the design of chips. A great deal of interesting and important research continues in model checking.

Objections, Myths, and Limitations

There are a number of "standard" objections to formal methods, for example, that they are too hard to use or expensive, hard to socialize, or that they require too much training. The rebuttals to these objections are straightforward, but situational. For example, how much is too much training? Certain organizations spend millions of dollars to attain certain capability maturity model (CMM) levels, but balk at investing significantly less money in training for the use of formal methods. In any case, be wary of out-of-hand dismissal of formal methods based on naïve points. Formal methods are not for every situation, but they are certainly not for "no" situation, and they really ought to be considered in many situations.

Objections and Myths

Some objections to using formal methods include the fact that they can be error prone (just as mathematical theorem proving or computer programming are) and that sometimes they are unrealistic for some systems. These objections are valid, sometimes. But formal methods are not intended to be used in isolation, nor do they take the place of testing. Formal methods are complementary to other quality assurance approaches. Some of the other "standard" objections to formal methods are based on myth. Papers by Hall, and Bowen and Hinchey help capture and rebut these misconceptions. The first set of myths is as follows (Hall 1990).

1. *Myth: Formal methods can guarantee that software is perfect.* Truth: Nothing can guarantee that software will be perfect. Formal methods are one of many techniques that improve software qualities, particularly reliability.
2. *Myth: Formal methods are all about program proving.* Truth: We have shown that formal methods involve more than just program proving and involve expressing requirements with precision and economy, requirements validation, and model checking.
3. *Myth: Formal methods are only useful for safety-critical systems.* Truth: Formal methods are useful anywhere that high-quality software is desired.
4. *Myth: Formal methods require highly trained mathematicians.* Truth: While mathematical training is very helpful in mastering the nuances of expression, using formal methods is really about learning one or another formal language and about learning how to use language precisely. Requirements engineering must be based on the precise use of language, formal or otherwise.
5. *Myth: Formal methods increase the cost of development.* Truth: While there are costs associated with implementing a formal methods program, there are

associated benefits. Those benefits include a reduction in necessary rework downstream in the software lifecycle—at a more expensive stage in the project to make corrections.

6. *Myth: Formal methods are unacceptable to users.* Truth: Users will accept formal methods if they understand the benefits that they impart.
7. *Myth: Formal methods are not used on real, large-scale systems.* Truth: Formal methods are used in very large systems, including high-profile projects at NASA.

Five years after Hall's myths, Jon Bowen and Mike Hinchey gave us "Seven More Myths of Formal Methods" (Bowen and Hinchey 1995):

1. *Myth: Formal methods delay the development process.* Truth: If properly incorporated into the software development lifecycle, the use of formal methods will not delay, but actually accelerate the development process.
2. *Myth: Formal methods lack tools.* Truth: We have already seen that there are a number of tools that support formal methods—editing tools, translation tools, compilers, model checkers, and so forth. The available tool set depends on the formal method used.
3. *Myth: Formal methods replace traditional engineering design methods.* Truth: Formal methods enhance and enrich traditional design.
4. *Myth: Formal methods only apply to software.* Truth: Many formal methods were developed for hardware systems, for example, the use of Petri Nets in digital design.
5. *Myth: Formal methods are unnecessary.* Truth: This statement is true only if high-integrity systems are unnecessary.
6. *Myth: Formal methods are not supported.* Truth: Many organizations use, evolve, and promote formal methods. Furthermore, see Myth 2.
7. *Myth: Formal methods people always use formal methods.* Truth: "Formal methods people" use whatever tools are needed. Often that means using formal methods, but it also means using informal and "traditional" techniques too.

Limitations of Formal Methods

As has been mentioned, no method can guarantee absolute correctness, safety, and so on, and formal methods have some limitations. For example, they do not yet offer good ways to reason about alternative designs or architectures. Formal specifications must be converted to a design, then a conventional implementation language. This translation process is subject to all the potential pitfalls of any programming effort. And notation evolution is a slow, but ongoing process in the formal methods community. Finally, as new and more powerful notations are developed, it can take many years from when a notation is created until it is adopted in industry.

Every technique has its shortcomings. These limitations, however, should not be a deterrent to using formal methods.

Final Advice

Our final advice on the use of formal methods comes from a well-known paper, "Ten Commandments of Formal Methods" (Bowen and Hinchey 1995).

1. Thou shalt choose the appropriate notation.

   Comment: Whether you use B, Z, VDM, PVS, or whatever, pick the formal notation that is most appropriate for the situation. We do not advocate one formal language or another. What is appropriate for a situation will depend largely on what the organization is most comfortable using and the supporting tools available. An organization is not typically fluent in multiple formal languages, however, and it is not recommended that a company changes notation from one situation to the next. Training in the use of a formal language has to be considered if there are not enough engineers capable of using the tool of choice.

2. Thou shalt formalize, but not overformalize.

   Comment: Everything in moderation (including moderation).

3. Thou shalt estimate costs.

   Comment: Modern project management practice requires cost estimation at every step of the way.

4. Thou shalt have a formal method guru on call.

   Comment: It is always recommended to have access to an "expert" in formal methods. That expert may be an in-house specialist or a consultant.

5. Thou shalt not abandon thy traditional development methods.

   Comment: It has been continuously noted that formal methods are augmentative in nature.

6. Thou shalt document sufficiently.

   Comment: Documentation is critical in all systems engineering activities.

7. Thou shalt not compromise thy quality standards.

   Comment: Any objection here?

8. Thou shalt not be dogmatic.

   Comment: Dogmatism is not the way to promote the benefits of formal methods. We have tried to be balanced in our approach to formal methods usage.

9. Thou shalt test, test, and test again.

   Comment: Formal methods do not replace testing, they are supplemental to testing. An appropriate, lifecycle testing program is essential to modern systems engineering.

10. Thou shalt reuse.

    Comment: Take advantage of the power of formal methods, and one of the greatest of these is confidence in reuse. A software module that has been formally validated (as well as tested in the traditional sense) will be a good candidate for reuse because it is trusted.

Exercises

6.1 Are customers more likely to feel confident if formal methods are explained to them and then used?

6.2 Where in the software development process lifecycle do formal methods provide the most benefit?

6.3 Rewrite the train station specification in another formal language, such as Z or VDM.

6.4 Conduct a consistency check for the requirements found in Section 8.2 of the Smart Home SRS (video entry).

6.5 Conduct a consistency check for the requirements found in Section 8.3 of the Smart Home SRS (video playback).

References

Awodey, S. (2005) *Category Theory*, Pittsburg: Carnegie Mellon University.

Böhm, C., and G. Jacopini (1966) Flow diagrams, turing machines, and languages with only two formation rules, *Communications of the ACM* 9(5): 366–371.

Bowen, J.P., and M.G. Hinchey (1995) Seven more myths of formal methods, *Software*, 12(4): 34–41.

Bowen, J.P., and M.G. Hinchey (1995) Ten commandments of formal methods, *Computer*, 28(4): 56–63.

Clarke, E.M., and J.M. Wing (1996) Formal methods: state of the art and future directions, *ACM Computing Surveys*, 28(4): 626–643.

Crow, J., and B. Di Vita (1998) Formalizing Space Shuttle software requirements: four case studies, *ACM Transactions on Software Engineering and Methodology*, 7(3).

Forster, T.E. (2003) *Logic, Induction and Sets*, Cambridge: Cambridge University Press.

Gries, D., and P. Gries (2005) *Multimedia Introduction to Programming Using Java*, Springer.

Hall, A. (1990) Seven myths of formal methods, *Software*, 7(5): 11–19.
Hinchey, M.G. (1993) Formal methods for system specification, *Potentials*, 12(3): 50–52.
Hoare, C.A.R. (1969) An axiomatic basis for computer programming, *Communications of the ACM*, 12(10): 576–583.
IEEE Std 610.12-1990 IEEE Standard Glossary of Software Engineering Terminology, IEEE Standards, Piscataway, NJ.
Lano, K. (1996) *The B Language and Method*, Springer-Verlag, pp. 94–98.
Truss, L. (2004) *Eats Shoots and Leaves: The Zero Tolerance Approach to Punctuation*, Profile Books.

Chapter 7

Requirements Specification and Agile Methodologies

Introduction to Agile Methodologies*

Agile methodologies are a family of nontraditional software development strategies that have captured the imagination of many who are leery of traditional, process-laden approaches. Agile methodologies are characterized by their lack of rigid process, though this fact does not mean that agile methodologies, when correctly employed, are not rigorous nor suitable for industrial applications—they are. What is characteristically missing from agile approaches, however, are "cookbook" approaches (for example, those prescribed in the Project Management Body of Knowledge) and are therefore sometimes called "lightweight" approaches.

Agile methodologies are traditionally applied to software engineering, and while there are elements that can be applied to the engineering of systems that are not pure software (in particular, the human considerations), they are not typically applied to pure hardware systems. This is so because agile methodologies depend on a series of rapid, non-throwaway prototypes, an approach that is not often practical in hardware-based systems. In any case, the non-software engineer can still benefit from this

* Some of this section has been excerpted from Phillip A. Laplante, *What Every Engineer Needs to Know About Software Engineering*, CRC/Taylor & Francis, 2006, with permission.

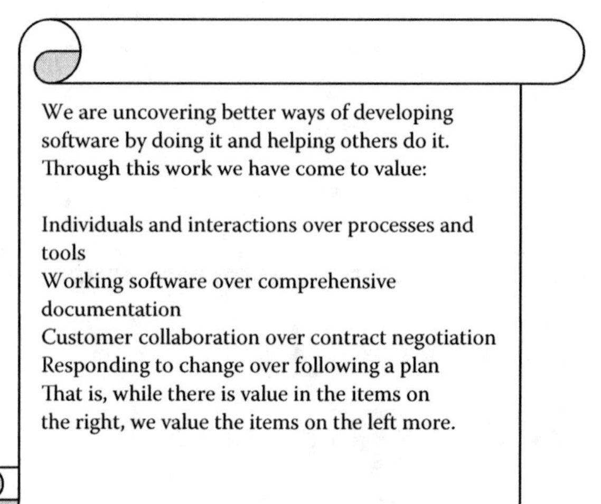

We are uncovering better ways of developing
software by doing it and helping others do it.
Through this work we have come to value:
Individuals and interactions over processes and
tools
Working software over comprehensive
documentation
Customer collaboration over contract negotiation
Responding to change over following a plan
That is, while there is value in the items on
the right, we value the items on the left more.

Figure 7.1 Manifesto for agile software development (Beck 2000).

chapter because agile methodologies are increasingly being employed, and because the mindset of the agile software engineer includes some healthy perspectives.

In order to fully understand the nature of agile methodologies, we need to examine a document called the Agile Manifesto and the principles behind it. The Agile Manifesto was introduced by a number of leading proponents of agile methodologies in order to explain their philosophy (see Figure 7.1).

Signatories to the Agile Manifesto include many luminaries of modern software engineering practice such as Kent Beck, Mike Beedle, Alistair Cockburn, Ward Cunningham, Martin Fowler, Jim Highsmith, Ron Jeffries, Brian Marick, Robert Martin, Steve Mellor, and Ken Schwaber. Underlying the Agile Manifesto is a set of principles. Look at the principles below, noting the emphasis on those aspects that focus on requirements engineering, which we set in bold.

Principles Behind the Agile Manifesto

- At regular intervals, the team reflects on how to become more effective, then tunes and adjusts its behavior accordingly.
- Our highest priority is to satisfy the customer through early and continuous delivery of valuable software.
- *Welcome changing requirements, even late in development.* Agile processes harness change for the customer's competitive advantage.
- Deliver working software frequently, from a couple of weeks to a couple of months, with a preference to the shorter timescale.
- *Business people and developers must work together daily throughout the project.*

- Build projects around motivated individuals. Give them the environment and support they need, and trust them to get the job done.
- *The most efficient and effective method of conveying information to and within a development team is face-to-face conversation.*
- Working software is the primary measure of progress.
- Agile processes promote sustainable development. The sponsors, developers, and users should be able to maintain a constant pace indefinitely.
- Continuous attention to technical excellence and good design enhances agility.
- Simplicity—the art of maximizing the amount of work not done—is essential. "Do the simplest thing that could possibly work."
- *The best architectures, requirements, and designs emerge from self-organizing teams.* (Beck 2000)

Notice how the principles acknowledge and embrace the notion that requirements change throughout the process. Also, the agile principles emphasize frequent, personal communication (this feature is beneficial in the engineering of non-software systems too). The highlighted features of requirements engineering in agile process models differ from the "traditional" waterfall and more modern models such as Iterative, Evolutionary, or Spiral development. These other models favor a great deal of up-front work on the requirements engineering process and the production of, often, voluminous requirements specifications documents.

Agile software development methods are a subset of iterative methods* that focus on embracing change and stress collaboration and early product delivery, while maintaining quality. Working code is considered the true artifact of the development process. Models, plans, and documentation are important and have their value, but exist only to support the development of working software, in contrast with the other approaches already discussed. However, this does not mean that an agile development approach is a free-for-all. There are very clear practices and principles that agile methodologists must embrace.

Agile methods are adaptive rather than predictive. This approach differs significantly from those models previously discussed, models that emphasize planning the software in great detail over a long period of time and for which significant changes in the software requirements specification can be problematic. Agile methods are a response to the common problem of constantly changing requirements that can bog down the more "ceremonial" up-front design approaches, which focus heavily on documentation at the start.

Agile methods are also "people-oriented" rather than process-oriented. This means they explicitly make a point of trying to make development "fun." Presumably,

* Most people define agile methodologies as being incremental. But incremental development implies that the features and schedule of each delivered version are planned. In some cases, agile methodologies tend to lead to versions with feature sets and delivery dates that are almost always not as planned.

this is because writing software requirements specifications and software design descriptions is onerous and hence, to be minimized.

Agile methodologies sometimes go by funny names like Crystal, Extreme Programming, and Scrum. Other agile methods include Dynamic Systems Development Method (DSDM), Feature-Driven Development, and adaptive programming, and there are many more. We will look more closely at two of these, XP and Scrum.

Extreme Programming (XP)

Extreme Programming (XP)* is one of the most widely used agile methodologies. XP is traditionally targeted towards smaller development teams and requires relatively few detailed artifacts. XP takes an iterative approach to its development cycles. We can visualize the difference in process between a traditional waterfall model, iterative models, and XP (Figure 7.2). Whereas an evolutionary or iterative method may still have distinct requirements analysis, design, implementation, and testing phases similar to the waterfall method, XP treats these activities as being interrelated and more-or-less-continuous.

XP promotes a set of twelve core practices that help developers to respond to and embrace inevitable change. The practices can be grouped according to four practice areas:

- planning
- coding
- designing
- testing

Some of the distinctive planning features of XP include holding daily stand-up meetings,† making frequent small releases, and moving people around. Coding practices include having the customer constantly available, coding the unit test cases first, and employing pair-programming (a unique coding strategy where two developers work on the same code together). Removal of the territorial ownership of any code unit is another feature of XP.

Design practices include looking for the simplest solutions first, avoiding too much planning for future growth (speculative generality), and refactoring the code (improving its structure) continuously.

---

* Extreme programming is sometimes also written "eXtreme Programming" to highlight the "XP."

† Stand up meetings are ten minute or less long where participants literally stand up and give an oral report of the previous day's activities and plans for that day.

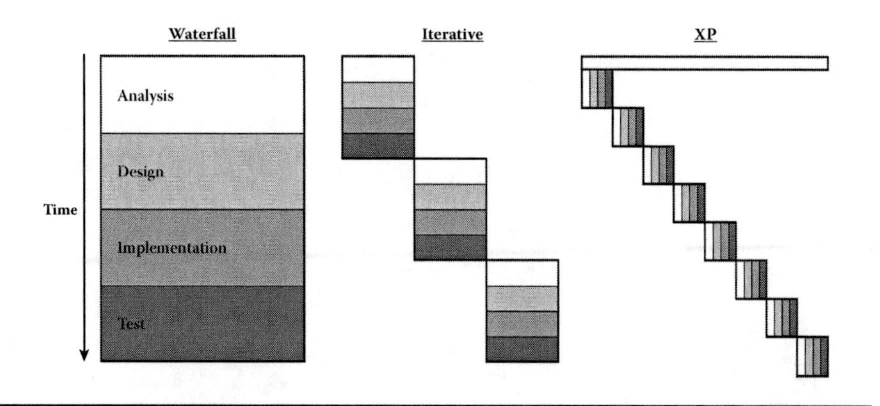

Figure 7.2 Comparison of waterfall, iterative, and XP development cycles (Beck 2000).

Testing practices include creating new test cases whenever a bug is found and having unit tests for all code, possibly using such frameworks as XUnit.

Scrum

Scrum, which is named after a particularly contentious point in a rugby match, enables self-organizing teams by encouraging verbal communication across all team members and across all stakeholders. The fundamental principle of Scrum is "empirical challenges cannot be addressed successfully in a traditional 'process control' manner," meaning that the problem cannot be fully understood or defined. Scrum encourages self-organization by fostering high-quality communication between all stakeholders. In this case it is implicit that the problem cannot be fully understood or defined (it may be a wicked problem). And the focus in Scrum is on maximizing the team's ability to respond in an agile manner to emerging challenges.

Scrum features a living backlog of prioritized work to be done. Completion of a largely fixed set of backlog items occurs in a series of short (approximately 30 days) iterations or sprints. Each day, a brief (e.g., 15-minute) meeting or Scrum is held in which progress is explained, upcoming work is described, and impediments are raised. A brief planning session occurs at the start of each sprint to define the backlog items to be completed. A brief post-mortem or heartbeat retrospective occurs at the end of the sprint (Figure 7.3).

A "ScrumMaster" removes obstacles or impediments to each sprint. The ScrumMaster is not the leader of the team (as they are self-organizing) but acts as a productivity buffer between the team and any destabilizing influences. In some organizations the role of the ScrumMaster can cause confusion. For example, if two members of a Scrum team are not working well together, it might be expected by a senior manager that the ScrumMaster "fix" the problem. Fixing team dysfunction

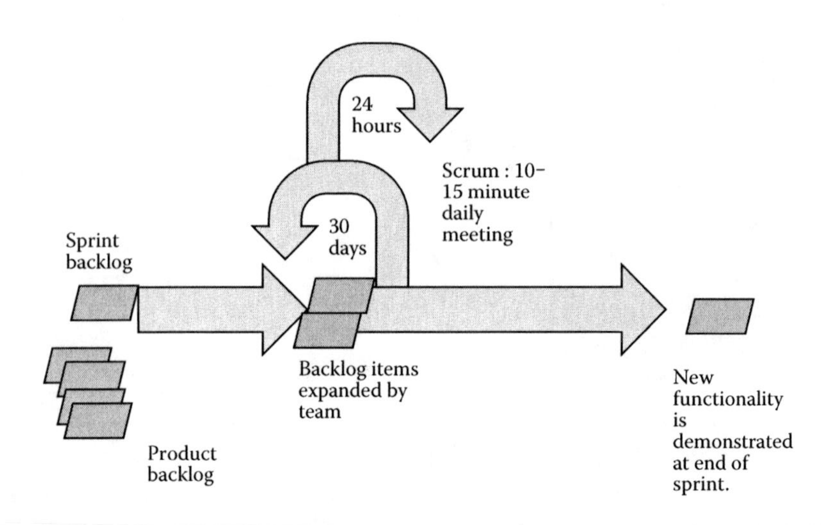

Figure 7.3 The Scrum development process from Boehm and Turner (2003), which itself was adapted from Schwaber and Beedle (2001).

is not the role of the ScrumMaster. Personnel problems need to be resolved by the line managers to which the involved parties report. This scenario illustrates the need for institution-wide education about agile methodologies when such approaches are going to be employed.

Requirements Engineering for Agile Methodologies

A major difference between traditional (here we mean, waterfall, evolutionary, iterative, spiral, etc.) and agile methodologies is in the gathering of requirements. In fact, some proponents of agile methodologies use the supposed advantages of agile requirements engineering practices as a selling point for agile methods. Requirements engineering approaches for agile methodologies tend to be much more informal.

Another difference is in the timing of the requirements engineering activities. In traditional systems and software engineering, requirements are gathered, analyzed, refined, etc. at the front end of the process. In agile methods, requirements engineering is an ongoing activity; that is, requirements are refined and discovered with each system build. Even in spiral methodologies, where prototyping is used for requirements refinement, requirements engineering occurs much less so downstream in the development process.

Customers are constantly involved in requirements discovery and refinement in agile methods. All systems developers are involved in the requirements engineering activity and each can and should have regular interaction with customers. In traditional approaches, the customer has less involvement once the requirements specification has been written and approved, and typically, the involvement is often not with the systems developers.

Presumably, the agile approach to requirements engineering is much more invulnerable to changes throughout the process (remember, "embrace change") than in traditional software engineering.

General Practices in Agile Methodologies

Cao and Ramesh (2008) studied 16 software development organizations that were using either XP or Scrum (or both) and uncovered seven requirements engineering practices that were "agile" in nature.

1. Face-to-face communications (over written specifications)
2. Iterative requirements engineering
3. Extreme prioritization (ongoing prioritization rather than once at the start, and prioritization is based primarily on business value)
4. Constant planning
5. Prototyping
6. Test-driven development
7. Reviews and tests

Some of these practices can be found in non-agile development (for example, test-driven development and prototyping), but these seven practices were consistent across all of the organizations studied.

As opposed to the fundamental software requirements specification, the fundamental artifact in agile methods is a stack of constantly evolving and refining requirements. These requirements are generally in the form of user stories. In any case, these requirements are generated by the customer and prioritized—the higher the level of detail in the requirement, the higher the priority. As new requirements are discovered, they are added to the stack and the stack is reshuffled in order to preserve prioritization (Figure 7.4).

There are no proscriptions on adding, changing, or removing requirements from the list at any time (which gives the customer tremendous freedom). Of course, once the system is built, or likely while is it is being built, the stack of user stories can be converted to a conventional software requirements specification for system maintenance and other conventional purposes.

Agile Requirements Best Practices

Scott Ambler (2007) suggests the following best practices for requirements engineering using agile methods. Many of the practices follow directly from the principles behind the Agile Manifesto.

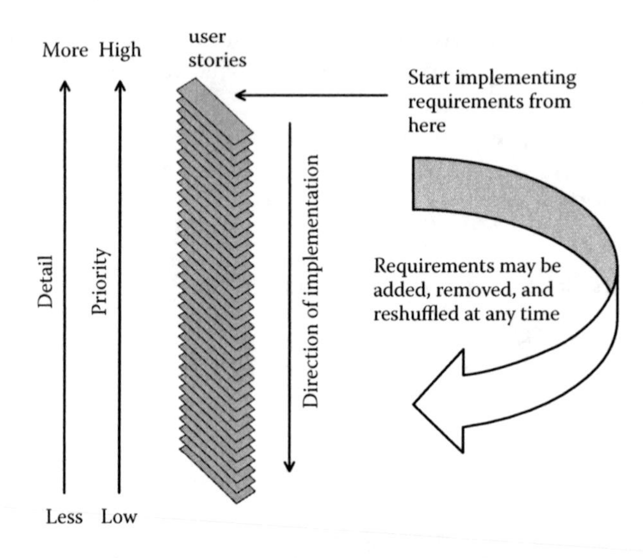

Figure 7.4 Agile requirements change management process adapted from http://www.agilemodeling.com/essays/agileRequirements.htm (Ambler 2007).

- Have active stakeholder participation.
- Use inclusive (stakeholder) models.
- Take a breadth-first approach.
- Model "storm" details (highly volatile requirements) just in time.
- Implement requirements, do not document them.
- Create platform-independent requirements to a point.
- Remember that "smaller is better."
- Question traceability.
- Explain the techniques.
- Adopt stakeholder terminology.
- Keep it fun.
- Obtain management support.
- Turn stakeholders into developers.
- Treat requirements like a prioritized stack.
- Have frequent, personal interaction.
- Make frequent delivery of software.
- Express requirements as features.

Ambler also suggests using such artifacts as CRCs, acceptance tests, business rule definitions, change cases, data flow diagrams, user interfaces, use cases, prototypes, features and scenarios, use cases diagrams, and user stories to model requirements (Ambler 2007). These elements can be added to the software requirements specification document along with the user stories.

For requirements elicitation he suggests using interviews (both in-person and electronic), focus groups, JAD, legacy code analysis, ethnographic observation, domain analysis, and having the customer on site at all times (Ambler 2007). For the remainder of this chapter our discussion focuses on the use of user stories to model requirements.

Requirements Engineering in XP

Requirements engineering in XP follows the model shown in Figure 7.4 where the stack of requirements in Ambler's model refers to user stories. And in XP, user stories are managed and implemented as code via the "planning" game.

The planning game in XP takes two forms: release and iteration planning. Release planning takes place after an initial set of user stories has been written. This set of stories is used to develop the overall project plan and plan for iterations. The set is also used to decide the approximate schedule for each user story and overall project.

Iteration planning is a period of time in which a set of user stories and fixes to failed tests from previous iterations are implemented. Each iteration is 1-3 weeks in duration. Tracking the rate of implementation of user stories from previous iterations (which is called "project velocity") helps to refine the development schedule.

Because requirements are constantly evolving during these processes, XP creator Kent Beck says that "in XP, requirements are a dialog, not a document" (Beck et al. 2001) although it is typical to convert the stack of user stories into a software requirements specification.

Requirements Engineering in Scrum

In Scrum, the requirements stack shown in the model of Figure 7.4 is, as in XP, the evolving backlog of user stories. And as in XP, these requirements are frozen at each iteration for development stability. In Scrum, each iteration takes about a month. To manage the changes in the stack, one person is given final authority for requirement prioritization (usually the product sponsor).

In Scrum the requirements backlog is organized into three types: product, release, and sprint. The product backlog contains the release backlogs, and each release contain sprint backlog. Figure 7.5 is a Venn diagram showing the containment relationship of the backlog items.

The product backlog acts as a repository for requirements targeted for release at some point. The requirements in the product backlog include low-, medium-, and high-level requirements.

The release backlog is a prioritized set of items drawn from the product backlog. The requirements in the release backlog may evolve so that they contain more details and low-level estimates.

Finally, the sprint backlog list is a set of release requirements that the team will complete (fully coded, tested, and documented) at the end of the sprint. These

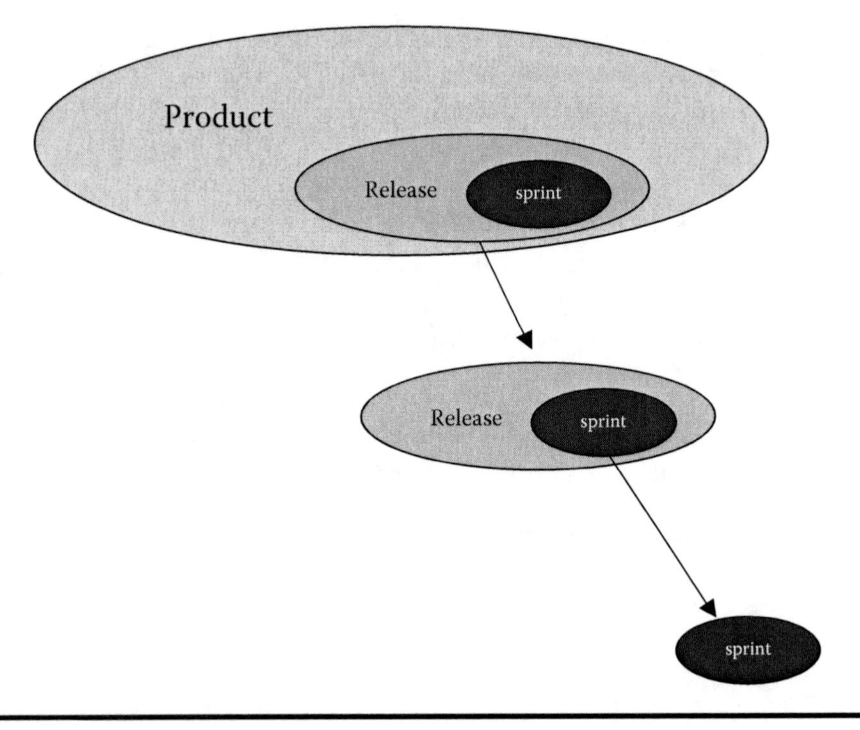

Figure 7.5 Backlog relationship between product, releases, and sprints.

requirements have evolved to a very high level of detail, and hence, their priority is high.

Scrum has been adopted in several major corporations, with notable success. Some of the author's students also use Scrum in courses. In these cases it proves highly effective when there is little time for long requirements discovery processes.

Writing User Stories

User stories are the most basic unit of requirements in most agile methodologies. Each user story represents a feature desired by the customer. User stories (a term coined by Kent Beck) are written by the customer on index cards, though the process can be automated via wikis or other tools. Formal requirements, use cases, and other artifacts are derived from the user stories by the software engineering team as needed.

A user story consists of the following components:

- Title—this is a short handle for the story. A present tense verb in active voice is desirable in the title.
- Acceptance test—this is a unique identifier that will be the name of a method to test the story.

- Priority—this is based on the prioritization scheme adopted. Priority can be assigned based on "traditional" prioritization of importance or on level of detail (higher priority is assigned to higher detail).
- Story points—this is the estimated time to implement the user story. This aspect makes user stories helpful for effort and cost estimation.
- Description—this is one to three sentences describing the story.

A sample layout for these elements on an index card is shown in Table 7.1.

Initial user stories are usually gathered in small offsite meetings. Stories can be generated either through goal-oriented (e.g., "let's discuss how a customer makes a purchase") approaches or through interactive (stream-of-consciousness) approaches. Developing user stories is an "iterative and interactive" process. The development team also manages the size of stories for uniformity (e.g., too large—split, too small—combine).

An example user story for a customer returning items in the pet store POS system is shown in Table 7.2.

User stories should be understandable to the customers and each story should add value.

Developers do not write user stories, users do. But stories need to be small enough that several can be completed per iteration. Stories should be independent (as much as possible); that is, a story should not refer back and forth to other stories. Finally, stories must be testable—like any requirement, if it cannot be tested, it's not a requirement! Testability of each story is considered by the development team.

Table 7.1 User Story Layout

<table>
<tr><td colspan="3" align="center">Title</td></tr>
<tr><td>Acceptance Test</td><td>Priority</td><td>Story Points</td></tr>
<tr><td colspan="3" align="center">Description</td></tr>
</table>

Table 7.2 User Story: Pet Store POS System

<table>
<tr><td colspan="3" align="center">Title: Customer returns items</td></tr>
<tr><td>Acceptance Test: custRetItem</td><td>Priority: 1</td><td>Story Points: 2</td></tr>
<tr><td colspan="3">When a customer returns an item, its purchase should be authenticated. If the purchase was authentic then the customer's account should be credited or the purchase amount returned. The inventory should be updated accordingly.</td></tr>
</table>

Table 7.3 User Story: Baggage Handling

Title: Detect Security Threat		
Acceptance Test: detSecThrt	*Priority: 1*	*Story Points: 3*
When a scanned bag has been determined to contain an instance of a banned item, the bag shall be diverted to the security checkpoint conveyor. The security manager shall be sent an email stating that a potential threat has been detected.		

Table 7.3 depicts another example user story describing a security threat detection in the airport baggage handling system.

Finally, it is worth noting that there is a significant difference between use cases and user stories. User stories come from the customer perspective and are simple and avoid implementation details. Use cases are more complex, and may include implementation details (e.g., fabricated objects). Customers don't usually write use cases (and if they do, beware, because now the customer is engaging in "software engineering"). Finally, it's hard to say what the equivalence is for the number of use cases per user story. One user story could equal one or more than 20 use cases. For example, for the customer return user story in Table 7.2, you can imagine that it will take many more use cases to deal with the various situations that can arise in a customer return. In agile methodologies user stories are much preferred to use cases.

Agile Requirements Engineering

We need to make a distinction between requirements engineering for agile methodologies and "agile requirements engineering." Agile requirements engineering means, generally, any ad hoc requirements engineering approach purported to be more flexible than "traditional" requirements engineering. This definition is not to be confused with specific practices for requirements engineering in agile methodologies as we just discussed (Sillitti, 2006).

A number of agile requirements engineering approaches have been introduced in the past few years, many of them not much more than sloppy requirements engineering. But some of the recent work in this area has been good too. However, for any "legitimate" agile requirements engineering methodologies that you may encounter, most of the practices can be traced to the Agile Manifesto.

To illustrate an "agile methodology," we describe one notable example. In this methodology called "Storytest-driven development" (SDD), many of the usual customer-facing features of agile methodologies are incorporated along with short bursts of iterative development à la XP or Scrum. One difference in SDD, however,

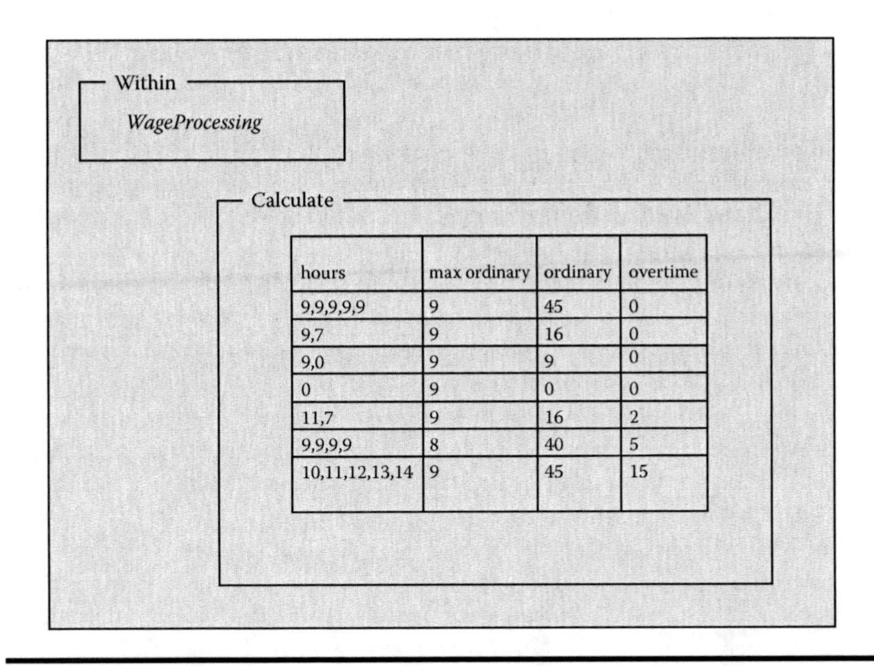

hours	max ordinary	ordinary	overtime
9,9,9,9,9	9	45	0
9,7	9	16	0
9,0	9	9	0
0	9	0	0
11,7	9	16	2
9,9,9,9	8	40	5
10,11,12,13,14	9	45	15

Figure 7.6 Using storytest to show how a company calculates pay for ordinary and overtime hours (Mugridge 2008).

is that instead of the conventional user stories, customers write or review "storytests." Storytests are nontechnical descriptions of behavior along with tests that verify that the behavior is correct. The storytests help "discover a lot of missing pieces or inconsistencies in a story" (Mugridge 2008).

A nice feature of the storytest is that it uses the Fit framework to allow users to build test cases into the stories in a tabular fashion (see Chapter 8). Hence, the users intuitively specify behavior and the tests for that behavior in the requirements.

For example, for a typical payroll system for a business (such as the pet store), a storytest describing how to calculate ordinary and overtime pay for an employee based on various clocked hours is given in Figure 7.6.

The table in Figure 7.6 shows, for instance, that if an employee works five consecutive nine-hour workdays, then they are considered to have 45 hours in ordinary and zero hours in overtime pay due. If a worker clocks 10, 11, 12, 13, and 14 hour days in a week, then they are entitled to 45 hours of regular and 15 hours of overtime pay. But the Fit framework is an interactive specification—plugging in different values into the table that are incorrect will show up as invalid (see Chapter 8).

Storytest-driven development is considered to be complementary to test-driven development (TDD) in that the former is applied to overall system development (Mugridge 2008).

Challenges for Requirements Engineering in Agile Methodologies

Of course, there are challenges in any new technology, and there are some in using agile methodologies, particularly with respect to requirements engineering. Laurie Williams (2004) discusses some of these shortcomings.

For example, agile methodologies do not always deal well with nonfunctional requirements. Why should this be so? One reason is that they are not always apparent when dealing with requirements functionality only (through user stories). Williams suggests dealing with this challenge by augmenting the user stories with appropriate nonfunctional requirements.

Another shortcoming is that with agile methodologies customer interaction is strong—but mostly through prototyping. As we have seen there are other ways to elicit nuanced requirements and understanding stakeholder needs, for example, using various interviewing techniques would be desirable.

Furthermore, with agile methodologies, validation ("is the system the right one?") is strong through testing and prototyping, but verification ("is the system correct?") is not so strong. She suggests that using formal methods could strengthen agile requirements engineering.

Finally, requirements management is built into the process (e.g., XP and Scrum), but it is mostly focused on the code level. Williams suggests that requirements management could be strengthened by adding more "standard" configuration management practices.

Exercises

7.1 How do you fit SRS documentation into an agile framework?

7.2 Is it possible to use agile methodologies when the customer is not on site. If so, how?

7.3 Why are agile methodologies generally not suitable for hardware-based projects as opposed to software projects?

7.4 Why can it be difficult for agile methodologies to cover nonfunctional requirements?

7.5 Are there any problems in encapsulating requirements into user stories?

7.6 For the pet store POS system generate a user story for customer purchases.

7.7 For the pet store POS system generate use cases for various customer purchases.

7.8 For the airport baggage handling system generate a user story for dealing with baggage that is to be diverted to another flight.

7.9 For the airport baggage handling system generate use cases for dealing with baggage that is to be diverted to another flight.

Bibliography

Ambler, S. http://www.agilemodeling.com/essays, last accessed 2007.

Beck, K. (2000) *Extreme Programming Explained: Embrace Change*, Longman Higher Education.

Beck, K., M. Beedle, A. van Bennekum, A. Cockburn, W. Cunningham, M. Fowler, J. Grenning, J. Highsmith, A. Hunt, R. Jeffries, J. Kern, B. Marick, R.C. Martin, S. Mellor, K. Schwaber, J. Sutherland, D. Thomas (2001) "Agile Manifesto" and "Principles Behind the Agile Manifesto." Online at http://agilemanifesto.org/, last accessed 12 January 2008.

Boehm, B., and R. Turner (2003) *Balancing Agility and Discipline: A Guide to the Perplexed*, Addison-Wesley.

Cao, L., and B. Ramesh (2008) Agile requirements engineering practices: An empirical study, *Software*, 25(1): 60–67.

Eberlein, A., and J.C. Sampaio do Prado Leite (2002) Agile Requirements International Workshop on Time-Constrained Requirements Engineering (TCRE'02), Essen, Germany, September 2002.

Laplante, P.A. (2006) *What Every Engineer Needs to Know About Software Engineering*, Boca Raton, FL: CRC/Taylor & Francis.

Mugridge, R. (2008) Managing agile project requirements with storytest-driven development, *Software*, 25(1): 68–75.

Schwaber, K., and M. Beedle (2001) *Agile Software Development with SCRUM*, Prentice Hall.

Sillitti, A., and G. Succi (2006) Requirements engineering for agile methods, in A. Aurum and C. Wohlin (Editors), *Engineering and Managing Software Requirements*, Springer, pp. 309–326.

Williams, L. (2004) *Agile Requirements Elicitation*. Online at http://ecommerce.ncsu.edu/studio/materials/AgileRE.pdf, last accessed 1 July, 2007.

Chapter 8

Tool Support for Requirements Engineering

Introduction

It has been estimated that by 2009 automated approaches to requirements engineering will reduce development costs by 30%. In addition, because of the use of these tools, user satisfaction with medium to large systems will improve from "fair" to "good" in that timeframe. Finally, by 2009, maintenance and enhancement costs for medium to large systems developed using such tools will decline by 10% (Light 2005).

Word processors, databases, spreadsheets, content analyzers, concept mapping programs, automated requirements checkers, and so on are all tools of interest to the requirements engineer. But the most celebrated requirements tools are large commercial packages that provide a high level of integrated functionality. The chief feature of these programs is the ability to represent and organize all "typical" requirements engineering objects such as use cases, scenarios, and user stories. Support for user-defined entities is also usually provided. Other typical functionalities for these large, commercial requirements engineering tools include

- multi-user support and version control
- online collaboration support
- customizable user interfaces
- built-in support for standards templates (such as IEEE 12207 and IEEE 830)

Table 8.1 Automated Requirements Engineering Tool Features (Heindl 2006)

Tool Feature	*Description*
Definition of workflow for requirements	A workflow (states, roles, state transitions) is configurable for requirements.
Automated generation of bi-directionality of traces	When the user creates a trace between artifact A and artifact B it automatically establishes a backward trace from B to A.
Definition of user-specific trace types	An authorized user can define trace types and assign names.
Suspect traces	When a requirement changes the tool automatically highlights all traces related to this requirement for checking and updating traces.
Long-term archiving functionality	All data in the tool can be archived in a format accessible without the tool if necessary.

- verification and validation tools
- customizable functionality through a programmable interface
- support for traceability
- user-defined glossary support (Heindl 2006)

We have already discussed automated requirements analysis via the NASA ARM tool in Chapter 5. And verification and validation features are an important component of any automated requirements engineering tool. Indeed, the more sophisticated commercial requirements engineering tools provide other requirements checking, tracing, and archiving features. These are shown in Table 8.1.

These features are particularly important because they provide for accurate tracing of artifacts over time. Traceability is an important characteristic of the SRS document and bears further discussion.

Traceability Support

Across the software lifecycle, traceability defines the relationships between requirements and their sources, design decisions, source code, and associated test cases. Within the SRS document, traceability focuses on the interrelationship between requirements and their sources, stakeholders, standards, regulations, and so on.

It is not unusual for one requirement to have an explicit reference to another requirement either as a "uses" or "refers to" relationship. Consider the following requirement:

> 2.1.1 The system shall provide for control of the heating system and be schedulable according to the time scheduling function (ref. requirement 3.2.2).

which depicts a "refers to" relationship. The "uses" relationship is illustrated in the following requirement:

> 2.1.1 The system shall provide for control of the heating system in accordance with the timetable shown in requirement 3.2.2 and the safety features described in requirement 4.1.

The main distinction between "uses" and "refers" to is that the "uses" represents a stronger, direct link.

The primary artifact of traceability is the requirements traceability matrix. The requirements traceability matrix can appear in tabular form in the SRS document, in a stand-alone traceability document, or internally represented within a tool for modification, display, and printout.

The entries in the requirements traceability matrix are defined as follows:

$R_{ij} = R$ if requirement i references requirement j (meaning "refers to" for informational purposes).

$R_{ij} = U$ if requirement i uses requirement j (meaning "depends on" directly).

R_{ij}= blank otherwise.

When a requirement both uses and references another requirement, the entry "U" supersedes that of "R." Since self-references are not included, the diagonal of the matrix always contains blank entries. We would also not expect circular referencing in the requirements so that if $R_{ij} = U$ or $R_{ij} = R$ we would expect R_{ji} to be blank. In fact, an automated verification feature in a requirements engineering tool should flag such circular references.

The format of a typical requirements traceability matrix is shown in Table 8.2.

A partial traceability matrix for a hypothetical system with sample entries is shown in Table 8.3.

Since R is usually sparse, it is convenient to list only those rows and columns that are non-blank. For example, for the SRS document for the Smart Home found in the appendix, the traceability matrix explicitly derived from the requirements is shown in Table 8.4.

Table 8.2 Format of Traceability Matrix (*R*) for Requirements Specification

Requirement ID	*1*	*1.1*	*1.1.1*	*…*	*1.2*	*1.2.1*	*…*	*2*	*2.1*	*2.1.1*	*…*
1											
1.1											
1.1.1											
…											
1.2											
1.2.1											
….											
2											
2.1											
2.1.1											
…											

Table 8.3 Partial Traceability Matrix for a Fictitious System Specification

Requirement ID	*1*	*1.1*	*1.1.1*	*1.1.2*	*1.2*	*1.2.1*	*1.2.2*	*2*	*2.1*	*2.1.1*	*3*
1									R	R	
1.1			U								
1.1.1	R							R			
1.1.2						R					
1.2										U	
1.2.1							R				
1.2.2										R	
2	U								R		
2.1				U							
2.1.1											R
3					R						

Table 8.4 Traceability Matrix for Smart Home Requirements Specification Shown in the Appendix

Requirement ID	*5.11*	*9.11*
9.1.1	R	
9.10.7		R

A sparse traceability matrix indicates a low level of explicit coupling between requirements. A low level of coupling within the requirements document is desirable—the more linked requirements the more changes to one requirement propagate to others. In fact, explicit requirements linkages are a violation of the principle of separation of concerns, a fundamental tenet of software engineering. Therefore, link requirements only when absolutely necessary.

Other kinds of linkages are desirable, however. For example, it is expected that each requirement will be linked to multiple tests. Test span metrics are used to characterize the linkages between the SRS document and the test plan in order to identify insufficient or excessive testing. The typical test span metrics are

- requirements per test
- tests per requirement

Research is still ongoing to determine appropriate ranges for these statistics. In any case, there is a tradeoff between time and cost of testing versus the comprehensiveness of testing. Planning for testing is part of the requirements engineering effort, as we have discussed exhaustively, and these linkages need to be anticipated in the SRS document.

Commercial Requirements Engineering Tools

While it is not our place to promote any particular commercial requirements engineering tools, it is appropriate to briefly survey the features of a few of the more commonly adopted ones. Major commercial requirements engineering tools that we will discuss include

- DOORS
- Rational RequisitePro
- Requirements and Traceability Management
- CaliberRM
- QFD/Capture

We will provide a very brief summary of their features.

DOORS

Telelogic's DOORS (Dynamic Object Oriented Requirements System) is essentially an object database for requirements management. That is, each feature is represented as an object containing a feature description, feature graph, and use case diagram. Feature attributes include mandatory, optional, single adaptor, and multiple adaptor, and the user can specify whether an attribute is required or excluded, which embodies the complementary and supporting nature of requirements. Other attributes are based on use case packages and product instances (Eriksson et al. 2005).

DOORS offers linking between all objects in a project for full traceability and missing link analysis. The tool can be customized via a C-like programming language. Standard templates are available to structure requirements in compliance with the ISO 12207, ISO 6592, and IEEE 830 software standards (Volere).

Rational RequisitePro

IBM's Rational RequisitePro is one of the most popular requirements and use case management tools. The tool integrates with Microsoft Word for easy requirements organization, integration, and analysis. There is also detailed attribute customization and filtering and a variety of traceability views that display parent/child relationships and highlight requirements that may be affected by later change.

The tool is capable of performing project-to-project comparisons using exportable XML-based files. Perhaps most importantly, the tool integrates with multiple tools in the IBM Software Development Platform (Volere).

Requirements and Traceability Management

Requirements and Traceability Management by Serena Software is an Oracle-based tool for management and traceability of requirements, designs, tests, project tasks, and other development information. The software provides a class definition tool used to model any type of hierarchical project data such as the SRS document, requirements hierarchies, system elements, and work breakdown structures. Generic relationships within the hierarchies can be established to cross-reference link information. The tool also provides an online collaboration feature (Volere).

CaliberRM

Borland's CaliberRM is another widely used requirements management tool. Features include a centralized repository for requirements artifacts, a customizable graphical user interface, full traceability across the lifecycle, and support for online glossaries (Volere).

QFD/Capture

QFD/Capture by International TechneGroup, Inc. provides basic support for the Quality Function Deployment approach. The tool provides predefined project templates, calculations and linkages between views of the "house of quality." In the tool, relationships can be defined in several ways, using direct input into a spreadsheet or through tree diagrams. Metrics for each requirement in a graphical tree and branch format are also supported. Requirements are linked so that as the project evolves changes in one area are reflected throughout the project (Volere).

Open Source Requirements Engineering Tools

As of this writing, there are more than 200,000 open source projects including all kinds of tools, utilities, libraries, and games. Not surprisingly, then, there are many open source solutions for requirements engineering, some as good as or almost as good as their commercial equivalents. We would like to take a look at three open source utilities that can be of use to the requirements engineer. This selection is a sampling only, and we urge you to turn first to open source repositories to look for tools before purchasing or trying to develop them from scratch.

FreeMind

FreeMind is an open source mind mapping software tool written in Java. FreeMind allows a user, stakeholder, or requirements engineer to organize ideas in a convenient, hierarchical diagram. The tool is quite useful for brainstorming, task analysis, laddering, traceability, JAD, focus groups, and many other requirements engineering activities (FreeMind).

The tool has an easy-to-use graphical user interface. Concepts are mapped in a hierarchical structure based on a primary concept and subordinate ideas to that concept. Handy icons can be used to mark priorities, important criteria, or hazards for each idea. Many other icons are available to enrich the content of the mind map.

To illustrate the use of the tool, let's look at Phil's concept of the Smart Home control system. Starting off with the basic concept of a "Smart Home" Phil creates a parent node (represented by a bubble) as shown in Figure 8.1.

Next, Phil uses the tool to attach a feature to the system (security) and assign a priority to that feature. Although there are up to seven levels of rankings available, Phil chose a simple three-level system.

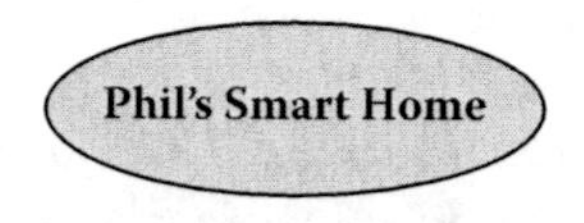

Figure 8.1 Phil's Smart Home system.

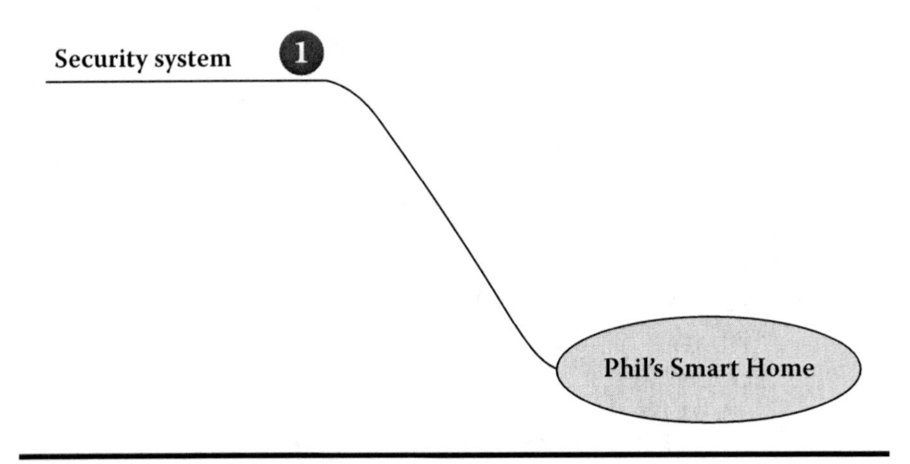

Figure 8.2 Adding the first feature to Phil's Smart Home system.

1. Mandatory
2. Optional
3. Nice to have

The resultant ranked feature and updated mind map is shown in Figure 8.2.

Phil then adds details to the feature. For example, he wants the security system of the Smart Home to use and work with the existing security system in the home. Another feature he desires is for the HVAC (heating ventilation and air-conditioning) system to integrate with the Smart Home system. In addition, because Phil has delicate plants, pets, and collectibles in the house, it is important that the temperature never exceed 100° Fahrenheit, so this hazard is marked with a bomb icon. The revised mind map is shown in Figure 8.3.

Phil continues with his brainstorming of features, adding music management, lawn care, and a telephone answering system. Some of the features include important details that are marked with the appropriate symbol (Figure 8.4).

As he sees the system materialize in a visual sense, Phil thinks of more features and is able to make the appropriate prioritization decisions. Phil adds to his mind map accordingly, as seen in Figure 8.5.

The mind map shown in Figure 8.5 is not complete. But this mind map can be used during elicitation discussions with requirements engineers and can be easily evolved over time.

Of course, there are other open source and commercial "mind mapping" tools that have similar functionality to FreeMind. We are simply suggesting that such a tool can be very valuable for many requirements elicitation activities.

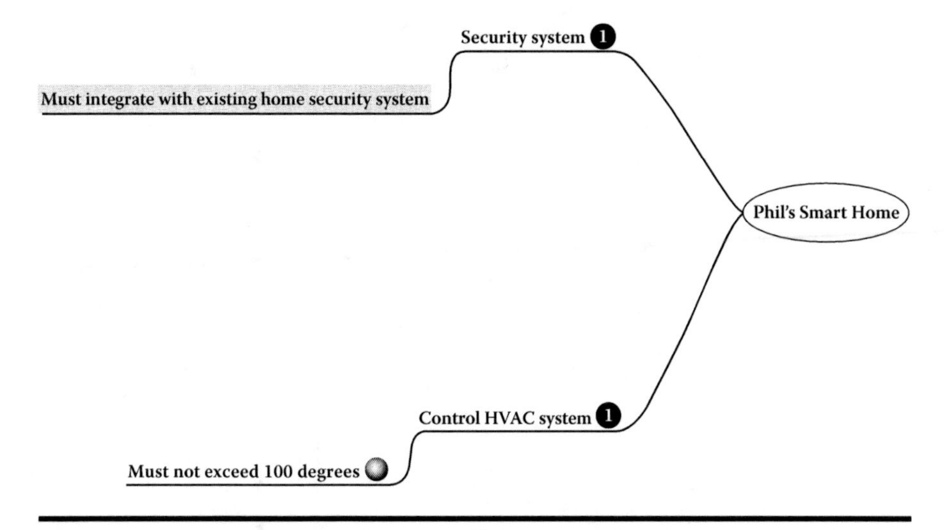

Figure 8.3 Adding a detail to the first feature to Phil's Smart Home system and adding a hazard.

Open Source Requirements Management Tool (OSRMT)

The Open Source Requirements Management Tool (OSRMT) is a platform-independent program written in Java and uses Swing and JBoss to provide full lifecycle support for requirements development and traceability. In OSRMT, features, requirements, design items, source code, and test cases are all input through a form-based GUI. Artifact details such as data dictionary elements, use case actions, test case steps, and so on are included in the forms. Each group of artifacts can be organized, sorted, searched, and filtered. User-defined artifacts and customized forms can also be specified.

The reporting feature allows for customizable reports based on the open source report writer Jasper. Plug-ins can be added for various supported databases such as Oracle, MySQL, SQL Server, and MS Access. Elements for reports can be exported to XML.

OSMRT is really a full-featured requirements engineering tool, and there is extensive online documentation for it. Therefore, it is beyond the scope to discuss installation and use of this open source tool. But the continued emergence of such free-for-use solutions may challenge commercial for-fee solutions and therefore cannot be ignored (OSRMT).

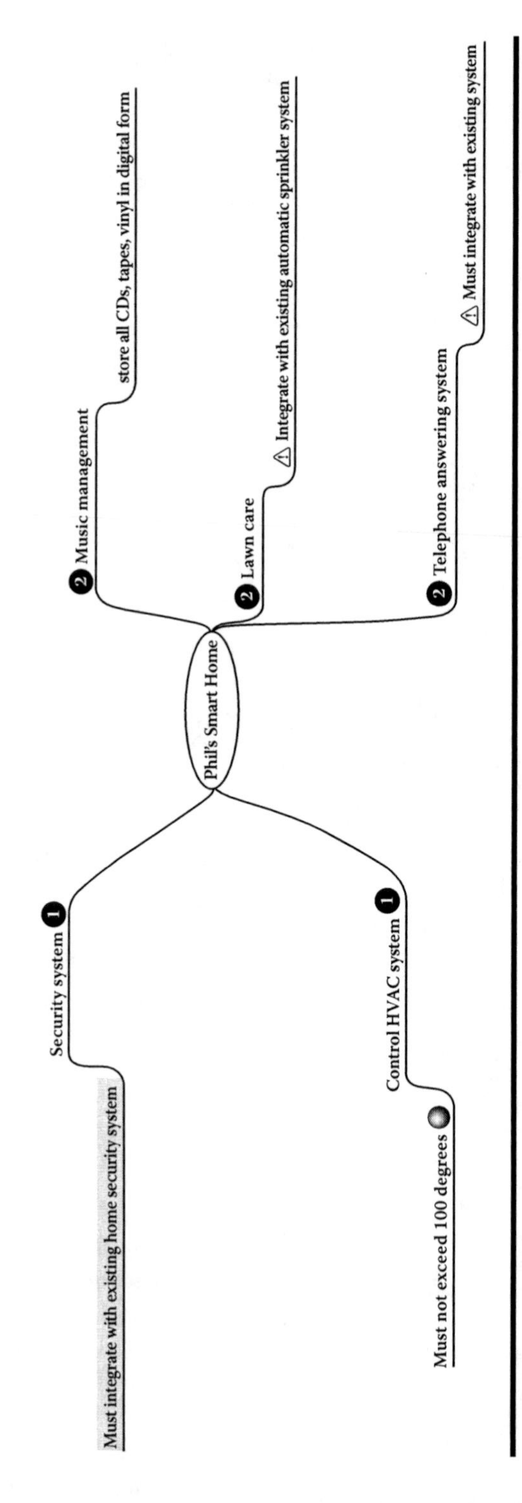

Figure 8.4 Adding more features to Phil's Smart Home system.

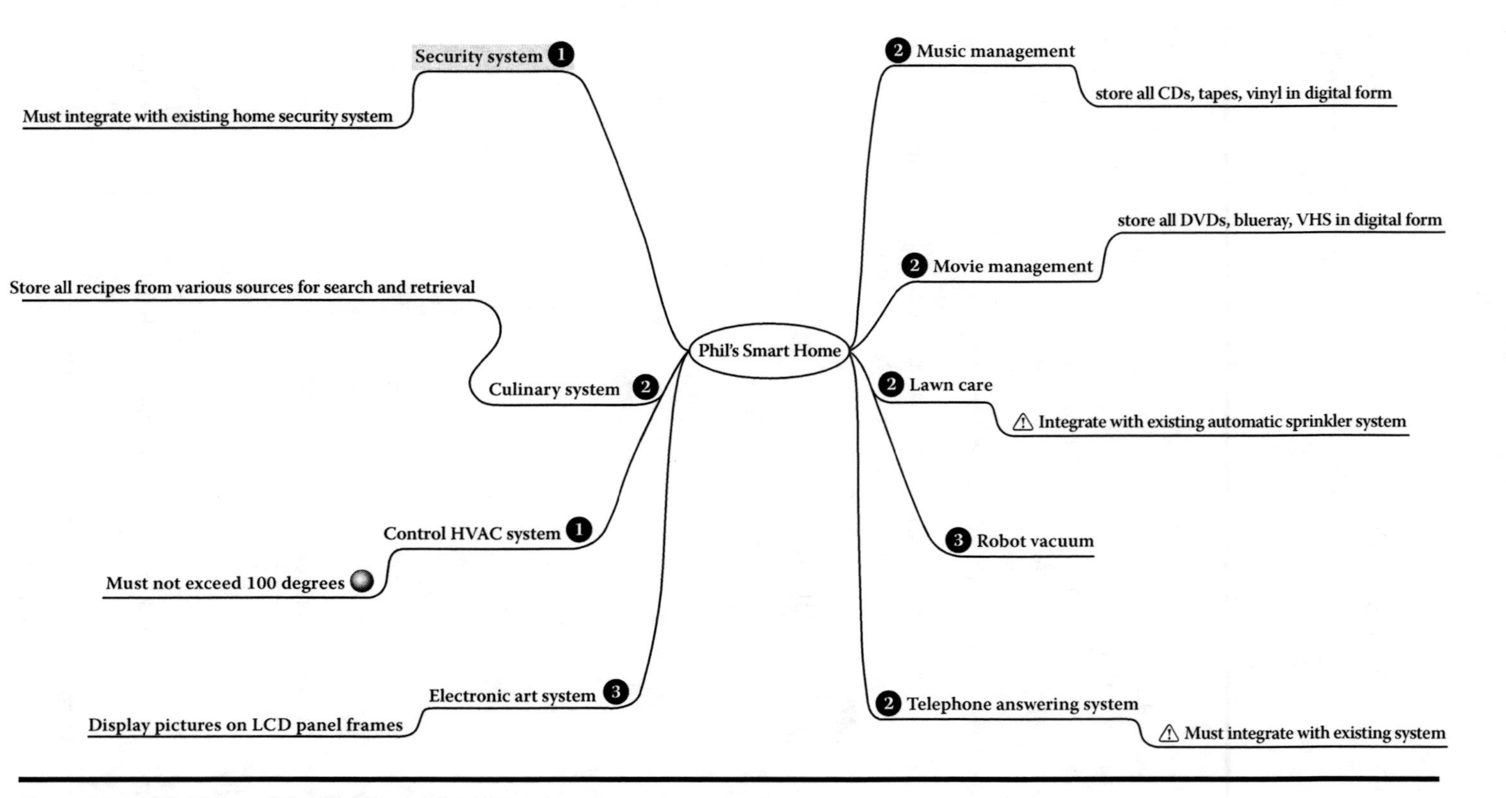

Figure 8.5 Phil's partial mind map of a Smart Home system.

FitNesse

FitNesse is an open source, Wiki-based software collaboration tool that provides a framework for building interactive test cases. While FitNesse was originally developed as a testing tool, our interest is in using it for integrated requirements/test specifications, for example, in an agile environment.

Before briefly describing how FitNesse works, it is appropriate to take a look at what a FitNesse test case looks like. The test case is, essentially, a table consisting of the name of the function (or method) to be tested, the set of input parameters and corresponding expected results parameters, and then a set of typical test cases across the rows of the table (see Table 8.5).

The FitNesse table is executable; when the actual code is developed, upon clicking a button, the results of the corresponding function will be calculated.

Let us illustrate the situation using the SRS for the Smart Home system in the appendix. In that SRS, Section 9.2.2 specifies that

> 9.2.2 System shall start coffee maker at any user defined time as long as water is present, coffee bean levels are sufficient and unit is powered.

Assuming that the preconditions are satisfied (coffee beans and water are available), a plausible schedule for making coffee might be as given in Table 8.6.

The schedule shown in Table 8.6 recognizes that the homeowner likes to sleep late on weekends. We could populate this table all different kinds of ways. Now it is true that the functionality desired is for dynamic scheduling so that, for example, during a holiday week, the home owner can sleep late every day. But the table here is used to indicate one scenario of appliance scheduling. In fact, if properly configured via the FitNesse framework, Table 8.6 constitutes an executable test case for the finished system.

Without going into too much detail, here is roughly how FitNesse works. FitNesse is a Wiki GUI built on top of the Fit framework. Fit runs the fixture code (a Java or C# class) corresponding to the test table. For example, in the top row of Table 8.5 "Activate_Coffee_Pot" specifies the actual class to be called. FitNesse colors the expected results cells red, green, or yellow, if the test failed, passed, or an exception was thrown, respectively (Gandhi 2005).

Table 8.5 Typical FitNesse Requirement/Test Case Format

Name of Function							
Input 1	*Input 2*	...	*Input n*	*Expected Result 1*	*Expected Result 2*	...	*Expected Result m*

Table 8.6 FitNesse Requirements Specification/Test Case for Activate Coffee Pot Function in Smart Home System

Activate_Coffee_Pot		
Day	*Time*	*Output*
Monday	7:00	coffee_pot_on_signal=high
Monday	8:00	coffee_pot_on_signal=low
Tuesday	7:00	coffee_pot_on_signal=high
Tuesday	8:00	coffee_pot_on_signal=low
Wednesday	7:00	coffee_pot_on_signal=high
Wednesday	8:00	coffee_pot_on_signal=low
Thursday	7:00	coffee_pot_on_signal=high
Thursday	8:00	coffee_pot_on_signal=low
Friday	7:00	coffee_pot_on_signal=high
Friday	8:00	coffee_pot_on_signal=low
Saturday	9:00	coffee_pot_on_signal=high
Saturday	10:00	coffee_pot_on_signal=low
Sunday	10:00	coffee_pot_on_signal=high
Sunday	11:00	coffee_pot_on_signal=low

Requirements Engineering Tool Best Practices

Whatever requirements engineering tool(s) you use, it is appropriate to use the tool judiciously and follow certain best practices. One set of such best practices is offered by Cleland-Huang et al. (2007):

Trace for a purpose. That is, determine which linkages are truly important; otherwise, a large number of extraneous links will be generated.

Define a suitable trace granularity. For example, linkages should be placed at the appropriate package, class, or method level.

Support in-place traceability. Provide traceability between elements as they reside in their native environments.

Use a well-defined project glossary. Create the glossary during initial discovery meetings with stakeholders and use consistently throughout the requirements engineering process.

Write quality requirements. Make sure to follow generally accepted best practices such as IEEE 830, which are particularly important for traceability.

Construct a meaningful hierarchy. Experimental results show that hierarchically organized requirements are more susceptible to intelligent linking software.

Bridge the intra-domain semantic gap. For example, avoid overloaded terminology, that is, words that mean completely different things in two different contexts.

Create rich content. Incorporate rationales and domain knowledge in each requirement.

Finally, be sure to use a process improvement plan for improving the requirements engineering process.

Exercises

8.1 What criteria should be used in choosing an appropriate requirements engineering tool?

8.2 Are there any drawbacks to using certain tools in requirements engineering activities?

8.3 When selecting an open source tool, what characteristics should you look for?

8.4 How can tools enable distributed, global requirements engineering activities? What are the drawbacks in this regard?

8.5 If an environment does not currently engage in solid requirements engineering practices, should tools be introduced?

8.6 What sort of problems might you find through a traceability matrix that you might not see without one?

8.7 Download FreeMind and use it to brainstorm a mind map for your Smart Home system.

8.8 Construct a FitNesse test table for the requirement described in Section 7.2 of the appendix.

References

Cleland-Huang, J., R. Settimi, E. Romanova, B. Berenbach, and S. Clark (2007) Best practices for automated traceability, *IEEE Computer*, June, pp. 27–35.

Eriksson, M., H. Morasi, J. Borstler, and K. Borg (2005) The PLUSS toolkit—Extending telelogic DOORS and IBM-Rational Rose to support product line in use case modeling, *Proceedings of the 20th IEEE/ACM international Conference on Automated software engineering* Long Beach, CA, USA, pp. 300–304.

FitNesse project website, www.fitnesse.org, last accessed 13 February 2008.

FreeMind project website http://freemind.sourceforge.net/wiki/index.php/Main_Page, last accessed 13 February 2008.

Gandhi, P. et al. (2005) Creating a living specification using Fit documents, *Agile 2005 Conf.* (Agile 05), IEEE CS Press, pp. 253–258.

Heindl, M., F. Reinish, S. Biffl, and A. Egyed (2006) Value-based selection of requirements engineering tool support, 32nd EUROMICRO Conference on Software Engineering and Advanced Applications. SEAA '06. Aug. 2006, pp. 266–273.

Light, M. (2005) Agile requirements definition and management will benefit application development. Online at www.cio.com/sponsors/GartnerReportonBenefitsofAgileRDM.PDF, last accessed 1 July 2007.

Open Source Requirements Management Tool open source community, http://www.osrmt.com, last accessed 14 February 2008.

Volere requirements resources. Online at http://www.volere.co.uk/tools.htm, last accessed 14 February 2008.

Chapter 9

Requirements Management

Introduction

Requirements management involves identifying, documenting, and tracking system requirements from inception through delivery. Inherent in this definition is understanding of the true meaning of the requirements and the management of customer (and stakeholder) expectations throughout the system's lifecycle. A solid requirements management process is the key to a successful project.

Most organizations do not have an explicit requirements management process in place, but this does not mean that requirements management does not occur within the organization. The requirements practices probably exist implicitly in the organization, but these practices are not usually documented. One of our recommendations, obviously, is to document the requirements management practices in your organization.

Managing Divergent Agendas

Each stakeholder has a different requirements "agenda." For example, business owners seek ways to get their money's worth from projects. Business partners want explicit requirements because they are like a "contract." Senior management expects more financial gain from projects than can be realized. And systems and software developers like uncertainty because it gives them freedom to innovate solutions.

Project managers may use the requirements to protect them from false accusations of underperformance in the delivered product.

One way to understand why the existence of different agendas—even among persons within the same stakeholder group—is the "Rashomon Effect." *Rashomon* is a highly revered 1950 Japanese film directed by Akira Kurosawa. The main plot involves the recounting of the murder of a samurai from the perspective of four witnesses to that event—the samurai, his wife, a bandit, and a wood cutter, each of whom has a hidden agenda, and tells a contradicting accounting of the event. Stated succinctly "your understanding of an event is influenced by many factors, such as your point of view and your interests in the outcome of the event" (Lawrence 1996).

The smart requirements manager seeks to manage these agendas by asking the right questions up front. Andriole (1998) suggests the following questions are appropriate:

1. What is the project request?
   - Who wants it?
   - Is it "discretionary" or "nondiscretionary?"
2. What is the project's purpose?
   - If completed, what impact will the new or enhanced system have on organizational performance?
   - On profitability?
   - On product development?
   - On customer retention and customer service?
3. What are the functional requirements?
   - What are the specific things the system should do to satisfy the purposeful requirements?
   - How should they be ranked?
   - What are the implementation risks?
4. What are the nonfunctional requirements, like security, usability, and interoperability?
   - How should they be ranked?
   - What are the implementation risks?
   - How do you trade off functional and nonfunctional requirements?
5. Do we understand the project well enough to prototype its functionality?
6. If the prototype is acceptable, will everyone sign off on the prioritized functionality and nonfunctionality to be delivered, on the initial cost and schedule estimates, on the estimates' inherent uncertainty, on the project's scope, and on the management of additional requirements?

Andriole asserts that by asking these questions up front, hidden agendas can be uncovered and differences resolved. At the very least, important issues will be raised up front and not much later in the process.

Expectation Revisited: Pascal's Wager

Mathematician Blaise Pascal (1623-1662) is well known for various achievements including Pascal's Triangle (a convenient way to find binomial coefficients) and work in probability. It was his work in probability theory that led to the notion of expected value, and he used such an approach later in life when he became more interested in religion than mathematics. It was during his monastic living that he developed a theory that suggested that it was advisable to live a virtuous life, whether or not one believed in a supreme being. His approach, using expected value theory, is now called Pascal's wager, and it goes like this.

Imagine an individual is having a trial of faith and is unsure if they believe in this supreme being (let's call him "God" for argument's sake) or not. Pascal suggests that it is valuable to consider the consequences of living virtuously, in the face of the eventual realization that such a God exists (or not). To see this, consider Table 9.1.

Assuming that it is equally probable that God exists or not (this is a big assumption), we see that the expected outcome (consequence) of living virtuously is half of Paradise while the expected outcome of living without virtue is half of Damnation. Therefore, it is in a person's best interests to live virtuously.

What does Pascal's wager have to do with expectation setting? Stakeholders will hedge their bets—sometimes withholding information or offering up inferior information because they are playing the odds involving various organizational issues. For example, does a stakeholder wish to request a feature that she believes no one else wants and for which she might be ridiculed? From a game theory standpoint it is safer for her to withhold her opinion. To see this, consider the modified Pascal's wager outcome matrix in Table 9.2.

If the stakeholder decides to speak out about a particular feature (or in opposition to a particular feature), assuming equi-likely probability that the group will

Table 9.1 Pascal's Wager Consequence Matrix

	God Exists	*God Does Not Exist*
Live virtuously	Achieve paradise	Null
Do not live virtuously	Achieve damnation	Null

Table 9.2 Modified Pascal's Wager Consequence Matrix

	Group Agrees	*Group Disagrees*
Speak out	Get praise	Get ridiculed
Remain silent	Nothing happens	Nothing happens

agree or disagree, the consequence matrix shows that she can expect to get some praise if the group agrees or some ridicule if the group disagrees. It is also well known that, in decision making, individuals will tend to make decisions that avoid loss over those that have the potential for gain—most individuals are risk averse. The foregoing analysis assumes that the probabilities of agreement and disagreement are the same—the expected consequences are much worse if the stakeholder believes there is a strong chance that her colleagues will disagree. Of course, later in the process the feature might suddenly be discovered by others to be important. Now it is very late in the game, however, and adding this feature is costly. Had the stakeholder only spoken up in the beginning, in a safe environment for discussion, a great deal of cost and trouble could have been avoided.

One last comment on Pascal's wager, expectation, and risk. The author once worked for a boss—we'll call him "Bob"—who greeted all new employees with a welcome lunch. At that lunch he would declare "I am a 'no surprises' kind of guy. You don't surprise me, and I won't surprise you. So, if there is ever anything on your mind, or any problem brewing, I want you to bring that to my attention right away." This sentiment sounded great. However, each time the author or anyone else would bring bad news to Bob, whether the messenger was responsible for the situation or not, Bob would blow his stack and berate the hapless do-gooder. After a while, potential messengers would forego bringing information to Bob. The decision was purely game theoretic—if you had bad news and you brought it to Bob, he would yell at you. If he didn't find out about it (as often happened because no one else would tell him, either), then you escaped his rampage. If he somehow found out about the bad news, you might get yelled at—but he might yell at the first person in his sight, not you, even if you were the party responsible for the problem. So, it made sense (and it was rigorously sound via game theory) to shut up.

It was rather ironic that "no surprise Bob" was always surprised because everyone was afraid to tell him anything. The lesson here is that, if you "shoot the messenger," people will begin to realize the consequences of bringing you information, and you will soon be deprived of that information. Actively seeking and inviting requirements information throughout the project lifecycle is an essential aspect of requirements management.

Global Requirements Management

Requirements engineering is one of the most collaboration-intensive activities in software development. Because of inadequate social contact, geographically distributed stakeholders and REs have trouble understanding requirements issues without the appropriate tools. Agile development seeks to solve this problem through always onsite customers (Sinha and Sengupta 2006).

Global and even onshore outsourcing presents all kinds of challenges to the requirements engineering endeavor. These include time delays and time zone issues, the costs and stresses of physical travel to client and vendor sites when needed, and the disadvantages of virtual conferencing and telephone. Even simple email communications cannot be relied upon entirely, even though for many globally distributed projects informal emails and email distributed documents are the main form of collaboration. But email use leads to frequent context switching, information fragmentation, and the loss of intuition communicated through nonverbal (or written) means.

When the offshoring takes place in a country with a different native language and substantially different culture, new problems may arise in terms of work schedules, work attitudes, communication barriers, and customer–vendor expectations of how to conduct business. Moreover, offshoring introduces a new risk factor: geopolitical risk—and this risk must be understood, quantified, and somehow factored into the requirements engineering process and schedule. Finally, there are vast differences in laws, legal process, and even the expectations of honesty in business transactions around the world. These issues are particularly relevant during the requirements elicitation phase.

Bhat et al. (2006) highlighted nine specific problems they observed or experienced. These included

- Conflicting client–vendor goals
- Low client involvement
- Conflicting requirements engineering approaches (between client and vendor)
- Misalignment of client commitment with project goals
- Disagreements in tool selection
- Communication issues
- Disowning responsibility
- Sign-off issues
- Tools misaligned with expectation

Bhat et al. suggest that the following success factors were missing in these cases, based on an analysis of their project experiences:

- Shared goal—that is, a project "metaphor"
- Shared culture—in the project sense, not in the sociological sense
- Shared process
- Shared responsibility
- Trust

These suggestions are largely consistent with agile methodologies, though we have already discussed the challenges and advantages of using agile approaches to requirements engineering.

Finally, what role can tools play in the globally distributed requirements engineering process? Sinha and Sengupta (2006) suggest that software tools can play an important role, though there are not many appropriate tools for this purpose. Appropriate software tools must support

- Informal collaboration
- Change management
- Promote awareness (e.g., auto-email stakeholders when triggers occur)
- Manage knowledge—provide a framework for saving and associating unstructured project information

There are several commercial and even open source solutions that claim to provide these features, but we leave the product research of these to the reader.

Antipatterns in Requirements Management*

In troubled organizations the main obstacle to success is frequently accurate problem identification. Diagnosing organizational dysfunction is quite important in dealing with the underlying problems that will lead to requirements engineering problems.

Conversely, when problems are correctly identified, they can almost always be dealt with appropriately. But organizational inertia frequently clouds the situation or makes it easier to do the wrong thing rather than the right thing. So how can you know what the right thing is if you've got the problem wrong?

In their groundbreaking book, Brown et al. (1998) described a taxonomy of problems or antipatterns that can occur in software architecture and design and in the management of software projects. They also described solutions or refactorings for these situations. The benefit of providing such a taxonomy is that it assists in the rapid and correct identification of problem situations, provides a playbook for addressing the problems, and provides some relief to the beleaguered employees in these situations in that they can take consolation in the fact that they are not alone.

These antipatterns bubble up from the individual manager through organizational dysfunction and can manifest in badly stated, incomplete, incorrect, or intentionally disruptive requirements.

Our antipattern set consists of an almost even split of 28 environmental (organizational) and 21 management antipatterns. *Management* antipatterns are caused by an individual manager or management team ("the management"). These antipatterns address issues in supervisors that lack the talent or temperament to lead a group, department, or organization. *Environmental* antipatterns are caused by a prevailing culture or social model. These antipatterns are the result of misguided corporate strategy or uncontrolled socio-political forces. But we choose to describe only a small subset of the antipatterns set that is particularly applicable in requirements engineering.

* Some of this discussion is excerpted from Laplante and Neill (2006) with permission.

Environmental Antipatterns

Divergent Goals

Everyone must pull in the same direction. There is no room for individual or hidden agendas that don't align with those of the business. The divergent goals antipattern exists when there are those who pull in different directions.

There are several direct and indirect problems with divergent goals.

- Hidden and personal agendas divergent to the mission of an organization starve resources from strategically important tasks.
- Organizations become fractured as cliques form to promote their own self-interests.
- Decisions are second-guessed and subject to "review by the replay official" as staff try to decipher genuine motives for edicts and changes.
- Strategic goals are hard enough to attain when everyone is working towards them, without complete support they become impossible and put risk to the organization.

There is a strong correspondence between stakeholder dissonance and divergent goals, so be very aware of the existence of both.

Since divergent goals can arise accidentally and intentionally there are two sets of solutions or refactorings.

Dealing with the first problem of comprehension and communication involves explaining the impact of day-to-day decisions on larger objectives. This was the case with the corporate executives of the box company described in Chapter 2. The executives forgot the bigger picture. There is more than providing a coffee mug with the mission statement, however. Remember that the misunderstanding is not because the staff are not aware of the mission or goals; organizations are generally very good at disseminating them. It is that they don't understand that their decisions have any impact on those goals. They have a very narrow perspective on the organization and that must be broadened.

The second problem of intentionally charting an opposing course is far more insidious, however, and requires considerable intervention and oversight. The starting point is to recognize the disconnect between their personal goals and those of the organization. Why do they feel that the stated goals are incorrect? If the motives really are personal, that they feel their personal success cannot come with success of the organization, radical changes are needed. Otherwise the best recourse is to get them to buy into the organizational goals. This is most easily achieved if every stakeholder is represented in the definition and dissemination of the core mission and goals, and subsequently kept informed, updated, and represented.

Process Clash

A process clash is the friction that can arise when advocates of different processes must work together without a proven hybrid process being defined. The dysfunction appears when organizations have two or more well-intended but non-complementary processes; a great deal of discomfort can be created for those involved. Symptoms of this antipattern include poor communications—even hostility—high turnover, and low productivity.

The solution to a process clash is as follows: develop a hybridized approach—one that resolves the differences at the processes' interfaces. Retraining and cross-training could also be used. For example, by training the analysis group in XP and the development group in RUP, better understanding can be achieved.

Another solution is to change to a third process that resolves the conflict. For example, domain driven modeling might have been used instead of RUP. Domain driven modeling can be used in conjunction with agile methodologies with no conflicts.

There is a strong correlation between the aforementioned two antipatterns, and certain capability maturity models, like the CMMI, can be used to help identify and reconcile process clashes that lead to divergent goals.

Management Antipatterns

Metric Abuse

The first management antipattern that might arise in requirements engineering is metric abuse, that is, the misuse of metrics either through incompetence or with deliberate malice (Dekkers and McQuaid 2002).

At the core of many process improvement efforts is the introduction of a measurement program. In fact sometimes the measurement program is the process improvement. That is to say, some people misunderstand the role measurement plays in management and misconstrue its mere presence as an improvement. This is not a correct assumption. When the data used in the metric are incorrect or the metric is measuring the wrong thing, the decisions made based upon them are likely the wrong ones and will do more harm than good.

Of course, the significant problems that can arise from metric abuse depend on the root of the problem: incompetence or malice.

- Incompetence: failure to understand the difference between causality and correlation; misinterpreting indirect measures; underestimating the effect of a measurement program.
- Malice: selecting metrics that support or decry a particular position based upon a personal agenda.

The solution or refactoring to the problem is as follows: The first thing to do is stop! Measuring nothing is better than measuring the wrong thing in many cases. When data are available, people use them in decision making, regardless of their accuracy.

Once the decks have been cleared Dekkers and McQuaid suggest a number of steps necessary for the introduction of a meaningful measurement program (Dekkers and McQuaid 2002):

1. *Define measurement objectives and plans*—perhaps by applying the goal-question-metric (GQM) paradigm.
2. *Make measurement part of the process*—don't treat it like another project that might get its budget cut or that one day you hope to complete.
3. *Gain a thorough understanding of measurement*—be sure you understand direct and indirect metrics; causality versus correlation; and, most importantly, that metrics must be interpreted and acted upon.
4. *Focus on cultural issues*—a measurement program will affect the organization's culture; expect it and plan for it.
5. *Create a safe environment to collect and report true data*—remember that without a good rationale people will be suspicious of new metrics, fearful of a time-and-motion study in sheep's clothing.
6. *Cultivate a predisposition to change*—the metrics will reveal deficiencies and inefficiencies so be ready to make improvements.
7. *Develop a complementary suite of measures*—responding to an individual metric in isolation can have negative side-effects. A suite of metrics lowers this risk.

If you believe that you are being metric mismanaged, then you can try to instigate the above process by questioning management about why the metrics are being collected, how they are being used, and whether there is any justification for such use. You can also offer to provide corrective understanding of the metrics with opportunities of alternate metrics and appropriate use or more appropriate uses of the existing metrics.

Mushroom Management

Mushroom management is a situation in which management fails to communicate effectively with staff. Essentially, information is deliberately withheld in order to keep everyone "fat, dumb, and happy." The name is derived from the fact that mushrooms thrive in darkness and dim light but will die in the sunshine. As the old saying goes "keep them in the dark, feed them dung, watch them grow … and then cut off their heads when you are done with them."

The dysfunction occurs when members of the team don't really understand the big picture; the effects can be significant, particularly with respect to requirements engineering when stakeholders get left out. It is somewhat insulting to assume that someone working on the front lines doesn't have a need to understand the bigger picture. Moreover, those who are working directly with customers, for example, might have excellent ideas that may have sweeping impact on the company. So, mushroom management can lead to low employee morale, turnover, missed opportunities, and general failure.

Those eager to perpetuate mushroom management will find excuses for not revealing information, strategy, and data. To refactor this situation some simple strategies to employ include

- Take ownership of problems that allow you to demand more transparency.
- Seek out information on your own. It's out there. You just have to work harder to find it and you may have to put together the pieces. Between you and the other mushrooms, you might be able to see most of the larger picture.
- Advocate for conversion to a culture of open-book management.

With all refactoring, courage and patience are needed to affect change.

Other Paradigms for Requirements Management

Requirements Management and Improvisational Comedy

Improvisational comedy can provide some techniques for collaboration (and after all requirements engineering is the ultimate in collaboration) and for dealing with adversity in requirements management. Anyone who has ever enjoyed improvisational comedy (for example, the television show *Whose Line Is It Anyway?*) has seen the exquisite interplay of persons with very different points of view and the resolution of those differences. The author has studied improvisational comedy and has observed a number of lessons that can be taken away from that art.

- Listening skills are really important—both to hear what customers and other stakeholders are saying and to play off your partner(s) in the requirements engineering effort.
- When there is disagreement or partial agreement the best response is "yes, and …" rather than "yes, but …" That is, build on, rather than tear down ideas.
- Things will go wrong—both improvisational comedy and requirements engineering are about adapting.
- You should have fun in the face of adversity.
- Finally, you should learn to react by controlling only that which is controllable (usually, it is only your own reaction to certain events, not the events themselves).

You can actually practice some techniques from improvisational comedy to help you develop your listening skills, emotional intelligence, and ability to think on your feet, which in turn, will improve your practice as a requirements engineer and as a team-player.

For example, consider one improvisational skill-building game called "Zip, Zap, Zup (or Zot)."* Here is how it works. Organize four or more people (the more, the better) to stand in a circle facing inside. One person starts off by saying either zip, zap, or zup. If that person looks at you, you look at someone else in the circle and reply in turn—zip, zap, or zup. Participants are allowed no other verbal communication than the three words. The "game" continues until all participants begin to anticipate which of the three responses is going to be given. The game is a lot harder to play than it seems, and the ability to anticipate responses can take several minutes (if it is attained at all). The point of this game/exercise is that it forces participants to "hear" emotions and pick up on other nonverbal cues.

In another game, "Dr. Know-it-all," a group of three or more participants answers questions together, with each participant providing just one word of the answer at a time. So, in a requirements engineering exercise, we would gather a collection of participants from stakeholder group A and ask them the following question. "Please complete the following sentence: the system should provide the following ..." then participants provide the answer one word each, in turn. This is a very difficult experience, and it is not intended as a requirements elicitation technique—it is a thought building and team building exercise and it can help to inform the requirements engineer and the participants about the challenges ahead.

One final exercise involves the requirements engineer answering questions from two other people at the same time. This experience helps the requirements engineer to think on their feet and also simulates what they will often experience when interacting with customers where they may need to respond simultaneously to questions from two different stakeholders/customers.

While this author prefers comedy as the appropriate genre for requirements elicitation, others (for example, Mayhaux and Maiden 2008) have studied improvisational theater as a paradigm for requirements discovery. In any case, it seems clear that our brains tend to suppress our best ideas and improvisation helps you to think spontaneously and generously. So, try these exercises for fun and to develop these important skills.

Requirements Management as Scriptwriting

In many ways the writing of screenplays for movies (scripts) has some similarities to requirements engineering. For example, Norden (2007) describes how requirements engineers can learn how to resolve different viewpoints by observing the screenwriting process. There are other similarities between the making of movies and software. These include

- Movies are announced well in advance while building certain expectations, which may not be delivered upon as the movie evolves (e.g., changes in actors,

* The author admits that this is also a traditional "drinking" game.

screenwriters, directors, plots, and release date). Software and systems are often announced in advance of its actual release and with subsequent changes in announced functionality and release date.

- Egos are often a significant factor in movies and software and systems.
- There are often too many hands involved in making movies and software systems.
- Sometimes the needs of movies exceed the technology (and new technology has to be developed as a result). The same is true for software systems.
- Movies often exceed budget and delivery expectations. Need we say more about software?
- Movies are filmed out of order, requirements are given, and software and some systems may be built this way, too.
- Movies are shot out of sequence and then assembled to make sense later. Software is almost always developed this way, too. Some systems are, too.
- A great deal of work ends up getting thrown away—in movies it ends up as film left on the cutting room floor, in software, as throwaway prototypes and one-time-use tools. While systems components are not usually built to be thrown away, many text fixtures are.

What can we learn from big picture production? In a short vignette within Norden's article, Sara Jones provides the following tips for requirements engineers from screenwriting:

- Preparation is everything. Don't leap straight into the detail, but do your homework first!
- Expect to work through many drafts. Remember that different versions might each need to support planning and enable collaboration between stakeholders.
- Think in detail about the future system users—their backgrounds, attitudes, habits, likes, and dislikes.
- Don't just think about the users' physical actions. Remember what they might think (their cognitive actions) and feel.
- Remember your requirements story's arc. Requirements should tell the complete story of achieving user goals.
- Hold your audience—remember that someone must read your requirements document. Perhaps there is a place for tension and dramatic irony. (Norden, 2007)

Reference Models for Requirements Management*

There are several different international standards that provide either reference process models, management philosophies, or quality standards that can be used

* Some of this section is excerpted from Phillip A. Laplante, *What Every Engineer Needs to Know About Software Engineering*, CRC/Taylor & Francis, 2006, with permission.

to inform the requirements engineering effort. These standards are not mutually exclusive in that they may be used in a complementary manner.

ISO 9000-3 (1997)

ISO Standard 9000 (International Standards Organization) is a generic, worldwide standard for quality improvement. The standard, which collectively is described in five standards, ISO 9000 through ISO 9004, was designed to be applied in a wide variety of manufacturing environments. ISO 9001 through ISO 9004 apply to enterprises according to the scope of their activities. ISO 9004 and ISO 9000-X family are documents that provide guidelines for specific applications domains. These ISO standards are process-oriented, "common-sense" practices that help companies create a quality environment.

While ISO 9000 is widely adopted in Europe an increasing number of U.S. and Asian companies have also adopted it.

Unfortunately these standards are very general and provide little specific process guidance. While certain recommendations are helpful as a "checklist," they provide very little in terms of a process that can be used. And even though a number of metrics have been made available to add some rigor to this somewhat generic standard, in order to achieve certification under the ISO standard, significant paper trails and overhead are required.

Six Sigma

Developed by Motorola, Six Sigma is a management philosophy based on removing process variation. Six Sigma focuses on the control of a process to ensure that outputs are within six standard deviations (six sigma) from the mean of the specified goals. Six Sigma is implemented using define, measure, analyze, improve, and control (DMAIC).

Define means to describe the process to be improved, usually through using some sort of business process model. Measure means to identify and capture relevant metrics for each aspect of the process model. The goal-question-metric paradigm is helpful in this regard.

Improve, obviously means to change some aspect of the process so that beneficial changes are seen in the associated metrics, usually by attacking the aspect that will have the highest payback.

Finally, analyze and control means to use ongoing monitoring of the metrics to continuously revisit the model, observe the metrics, and refine the process as needed.

Six Sigma is more process yield-based than CMMI so CMMI process areas can be used to support DMIAC (e.g., by encouraging measurement). And, while CMMI identifies activities, Six Sigma helps optimize those activities. Six Sigma can also provide specific tools for implementing CMMI practices (e.g., estimation and risk management).

Some organizations use Six Sigma as part of their software quality practice. The issue here, however, is in finding an appropriate business process model for the software production process that does not devolve into a simple, and highly artificial, waterfall process. If appropriate metrics for requirements can be determined (e.g., 830) then Six Sigma can be used to improve the RE process.

Capability Maturity Model (CMMI)

The Capability Maturity Model Integrative (CMMI) is a systems and software quality model consisting of five levels. The CMMI is not a lifecycle model, but rather a system for describing the principles and practices underlying process maturity. CMMI is intended to help software organizations improve the maturity of their processes in terms of an evolutionary path from ad hoc, chaotic processes to mature, disciplined processes.

Developed by the Software Engineering Institute at Carnegie Mellon University, the CMMI is organized into five maturity levels. Predictability, effectiveness, and control of an organization's software processes are believed to improve as the organization moves up these five levels. While not truly rigorous, there is some empirical evidence that supports this position.

The Capability Maturity Model describes both high-level and low-level requirements management processes. The high-level processes apply to managers and team leaders, while the low-level processes pertain to analysts, designers, developers, and testers.

Typical high-level requirements practices/processes include

- adhere to organizational policies
- track documented project plans
- allocate adequate resources
- assign responsibility and authority
- train appropriate personnel
- place all items under version or configuration control and have them reviewed by all (available) stakeholders
- comply with relevant standards
- review status with higher management

And low-level best practices/processes include

- understand requirements
- get all participants to commit to requirements
- manage requirements changes throughout the lifecycle
- manage requirements traceability (forwards and backwards)
- identify and correct inconsistencies between project plans and requirements

Achieving level 3 and higher for the CMM requires that these best practices be documented and followed within an organization.

IEEE 830

We have already discussed IEEE standard 830 in the context of risk mitigation. In terms of managing the requirements engineering activities, standard 830 provides two major things. First, the standard describes the qualities that govern good software requirements specifications. In addition, 830 provides a framework for approaching the organization of requirements (e.g., object-oriented, hierarchical, etc.).

From requirements management perspective, the "sample table of contents" is the least important offering of 830.

IEEE 12207 (2002)

ISO 12207: Standard for Information Technology—Software Life Cycle Processes, describes five "primary processes"—acquisition, supply, development, maintenance, and operation. ISO 12207 divides the five processes into "activities," and the activities into "tasks," while placing requirements upon their execution. It also specifies eight "supporting processes"—documentation, configuration management, quality assurance, verification, validation, joint review, audit, and problem resolution—as well as four "organizational processes"—management, infrastructure, improvement, and training.

The ISO standard intends for organizations to tailor these processes to fit the scope of their particular projects by deleting all inapplicable activities, and it defines 12207 compliance as being in terms of tailored performance of those processes, activities, and tasks.

12207 provides a structure of processes using mutually accepted terminology, rather than dictating a particular lifecycle model or software development method. Since it is a relatively high-level document, 12207 does not specify the details of how to perform the activities and tasks comprising the processes. Nor does it prescribe the name, format, or content of documentation. Therefore, organizations seeking to apply "12207" need to use additional standards or procedures that specify those details.

The IEEE recognizes this standard with the equivalent numbering: "IEEE/EIA 12207.0-1996, IEEE/EIA Standard Industry Implementation of International Standard ISO/IEC12207:1995, and (ISO/IEC 12207) Standard for Information Technology—Software Life Cycle Processes."

ISO/IEC 25030

We have already discussed this standard, which is complementary to IEEE 830, in Chapter 4.

A Case Study: FBI Virtual Case File

We close this chapter with a brief case study involving the U.S. Federal Bureau of Investigations disastrous Virtual Case File system, which was a failure of requirements management, among other things. This discussion is largely based on Goldstein's report of the project (2005). Begun in 2000, the Virtual Case File (VCF) was supposed to automate the FBI's paper-based work environment and allow agents and intelligence analysts to share investigative information. The system would also replace the obsolete Automated Case Support (ACS), which was based on 1970s software technology. But the five years and $170 million the project spent yielded only 700,000 lines of "bug-ridden" and dysfunctional code ($105 million worth unusable). Today the Virtual Case Project stands in a state of "limbo" (Alfonsi 2005).

What went wrong with VCF? Quite a bit went wrong, some of it political, some of it due to the complexities of a mammoth bureaucracy, but some of the failures can be attributed to bad requirements engineering. Goldstein notes that the project was poorly defined and that it suffered from slowly evolving requirements (800-page document to start). And even though JAD sessions were used for requirements discovery (several two-week sessions over six months), these were used late in the project and only after a change of vendors. Apparently, there was no discipline to stop requirements from snowballing. And an ever-changing set of sponsors (changes in leadership at the FBI as well as turnover in key congressional supporters) caused requirements drift.

Moreover, apparently, there was too much design detail in SRS, literally describing the color and position of buttons and text. And many requirements failed the simple 830 rules—often they were not "clear, precise, and complete."

Finally, there were also cultural issues at the FBI—sharing ideas and challenging decisions are not the norm. All of these factors, in any environment, would jeopardize the effectiveness of any requirements engineering effort.

A subsequent review of the project noted that there was no single person or group to blame—this was a failure on every level. But an independent fact-finding committee singled out FBI leadership and the prime contractor as being at fault for the reasons previously mentioned (Goldstein 2005).

Aside from the political ramifications, and the usual problems when designing large complex systems, what can be learned from the failure of VCF? From a requirements engineering standpoint, there are three very specific lessons. First, don't rush to get a system out before elicitation is completed. Obviously, there are pressures to deliver, particularly for such an expensive and high-profile system, but these pressures need to be resisted by strong leadership. Second, the system needs to be defined "completely and correctly" from the beginning. While this was highly problematic for the VCF due to the various changes in FBI leadership during the course of its evolution, a well-disciplined process could have helped to provide pushback against late requirements changes.

Finally, for very large systems it is helpful to have an architecture in mind. We realize that it has been said that the architecture should not be explicitly incorporated in the requirements specification, but having an architectural target can help inform the elicitation process, even if the imagined architecture is not the final one.

Exercises

9.1 Should a request to add or change features be anonymous?
9.2 How could metrics abuse begin to develop in an organization?
9.3 Give an example of process clash, from your own experience, if possible.
9.4 Give an example of metrics abuse, from your own experience, if possible.
9.5 Give an example of divergent goals, from your own experience, if possible.
9.6 How can CMMI be used to identify and reconcile process clash?

References

Alfonsi, B. (2005) FBI's virtual case file living in limbo, *Security & Privacy*, 3(2): 26–31.

Andriole, S. (1998) The politics of requirements management, *IEEE Software*, November/December, pp. 82–84.

Bhat, J.M., M. Gupta, and S.N. Murthy (2006) Overcoming requirements engineering challenges: Lessons from offshore outsourcing, *IEEE Software*, September/October, pp. 38–44.

Brown, W.J., R.C. Malveau, H.W. McCormick, and T.J. Mowbray (1998) *AntiPatterns: Refactoring Software, Architectures, and Projects in Crisis*, John Wiley and Sons.

Dekkers, C.A., and P.A. McQuaid (2002) The dangers of using software metrics to (mis) manage, *IT Professional*, 4(2): 24–30.

Goldstein, H. (2005) Who killed the virtual case file? *Spectrum*, 42(9): 24–35.

Laplante, P.A., and C.J. Neill (2006) *Antipatterns: Identification, Refactoring, and Management*, Auerbach Press.

Lawrence, B. (1996) Unresolved ambiguity, *American Programmer*, 9(5): 17–22.

Mayhaux, M., and N. Maiden (2008) Theater improvisers know the requirements game, *Software*, September/October, pp. 68–69.

Norden, B. (2007) Screenwriting for requirements engineers, *Software*, July, pp. 26–27.

Sinha, V., and B. Sengupta (2006) Enabling collaboration in distributed requirements management, *Software*, September/October, pp. 52–61.

Chapter 10

Value Engineering of Requirements

What, Why, When, and How of Value Engineering?

What Is Value Engineering?

Up until this point we have had very little discussion of the costs of certain features that customers may wish to include in the system requirements. Part of the reason is that it made sense to separate that very complex issue from all of the other problems surrounding requirements elicitation, analysis, representation, validation, verification, and so on. Another reason for avoiding the issue is that it is a tricky one.

There is a fundamental relationship between time, cost, and functionality. Project managers sometimes refer to this triad as the three legs of the project management stool. That is, you can't tip one without affecting the others. For example, you have already finished writing and agreeing to the specification and have provided an estimated cost to complete in response to some RFP (request for proposal). If the customer asks you to incorporate more features into the system, shouldn't the price and estimated time to complete increase? If not, then you were likely padding the original proposal, and the customer will not like that. Conversely, if you propose to build an agreed-upon system for a price, say $1 million, and the customer balks, should you then respond by giving her a lower price, say $800,000? Of course not, because if you lower the price without reducing the feature set or increasing the time to complete, the customer will think that you gave her an inflated price originally. The correct response, in this case, would be to agree to build the system for $800,000, but with fewer features (or taking much longer time).

To properly manage customer expectations, deal with tradeoffs between functionality, time, and cost, it is therefore necessary to have some way of estimating costs for system features. Making such estimations is not easy to do accurately at the early stages of a project, such as during the requirements engineering activities. However, it is necessary to make such cost and effort estimations. The activities related to managing expectations and estimating and managing costs are called value engineering.

When Does Value Engineering Occur?

Value engineering occurs throughout the system lifecycle and is typically considered a project management activity. But for the requirements engineer value engineering has to take place to help manage customer expectations concerning the final costs of the delivered system and the feasibility or infeasibility of delivering certain features.

During the early stages of elicitation, it is probably worthwhile to entertain some value engineering activities—but not too much. For example, you may wish to temper a customer's expectations about the possibility of delivering an elaborate feature if it will break the budget. On the other hand, if you are too harsh in providing cost assessments early, you can curtail the customer's creativity and frustrate them. Therefore, be cautious when conducting value engineering when working with user-level requirements.

The best time to conduct the cost analysis is at the time when the systems-level requirements are being put together. It is at this time that better cost estimates are available, and this is a time at which tradeoff decisions can be discussed more successfully with the customer, using some of the negotiating principles previously mentioned.

In the following sections, we'll look at some simple approaches to assist in the value engineering activity for requirements engineers. These sections provide only an introduction to these very complex activities that marry project management, cost accounting, and system engineering. Really, a set of experts with these skills is needed to provide the most accurate information for decision making.

Estimating Using COCOMO and Its Derivatives*

One of the most widely used software cost and effort estimation tools is Boehm's COCOMO model, first introduced in 1981. COCOMO is an acronym for con-

* This section is adapted from Phillip A. Laplante, *Software Engineering for Image Processing Systems*, CRC Press, September 2003, with permission.

structive cost model, which means that estimates are determined from a set of parameters that characterize the project. There are several versions of COCOMO including the original (basic), intermediate, and advanced models, each with increasing numbers of variables for tuning the estimates. The latest COCOMO models better accommodate more expressive modern languages as well as software generation tools that tend to produce more code with essentially the same effort. There are also COCOMO derivatives that are applicable for Web-based applications and software-intensive (but not pure software) systems.

COCOMO

COCOMO models are based on an estimate of lines of code, modified by a number of factors. The equations for project effort and duration are

$$\text{Effort} = A\prod_{i=1}^{9} cd_i (size)^{P_1} \tag{10.1}$$

$$\text{Duration} = B(Effort)^{P_2} \tag{10.2}$$

where A and B are a function of the type of software system to be constructed. For example, if the system is organic—that is, one that is not heavily embedded in the hardware—then the following parameters are used: $A = 3.2$, $B = 1.05$. If the system is semi-detached, that is, partially embedded, then these parameters are used: $A = 3.0$, $B = 1.12$. Finally, if the system is embedded, that is, closely tied to the underlying hardware like the visual inspection system, then the following parameters are used: $A = 2.8$, $B = 1.20$. Note that the exponent for the embedded system is the highest, leading to the longest time to complete for an equivalent number of delivered source instructions.

P_1 and P_2 are dependent on characteristics of the application domain and the cd_i are cost drivers based on a number of factors including

- product reliability and complexity
- platform difficulty
- personnel capabilities
- personnel experience
- facilities
- schedule constraints
- degree of planned reuse
- process efficiency and maturity
- precedentedness (that is, novelty of the project)
- development flexibility

- architectural cohesion
- team cohesion

with qualitative ratings on a Likert scale ranging from very low to very high, that is, numerical values are assigned to each of the responses.

Incidentally, effort represents the total project effort in person-months, and duration represents calendar months. These figures are necessary to convert the COCOMO estimate to an actual cost for the project.

In the advanced COCOMO models, a further adaptation adjustment factor is made for the proportion of code that is to be used in the system, namely, design modified, code modified, and integration modified. The adaptation factor, A, is given by Equation 10.3.

$$\begin{aligned} A = 0.4 \text{ (\% design modified)} + .03 \text{ (\% code modified)} \\ + 0.3 \text{ (\% integration modified)} \end{aligned} \tag{10.3}$$

COCOMO is widely recognized and respected as a software project management tool. It is useful even if the underlying model is not really understood. COCOMO software is commercially available and can even be found on the Web for free use.

An important consideration for the requirements engineer, however, is that COCOMO bases its estimation on lines of code, which are not easily estimated at the time of requirements engineering. Other techniques, such as function points, are needed to provide line of code estimates based on feature sets that are available when developing the requirements. We shall look at function points shortly.

WEBMO

WEBMO is a derivative of COCOMO II that is intended specifically for project estimation of Web-based projects, where COCOMO is not always as good. WEBMO uses the same effort and duration equations as COCOMO, but is based on a different set of predictors, namely

- number of function points
- number of xml, html, and query language links
- number of multimedia files
- number of scripts
- number of Web building blocks

with qualitative ratings on a Likert scale ranging from very low to very high and numerical equivalents shown in Table 10.1 (Reifer 2002).

Table 10.1 WEBMO Cost Drivers and Their Values (Reifer 2002)

	Ratings				
	Very Low	*Low*	*Nominal*	*High*	*Very High*
Cost Driver					
Product reliability	0.63 0	.85	1.0	1.30	1.67
Platform difficulty	0.75	0.87	1.00	1.21	1.41
Personnel capabilities	1.55	1.35	1.00	0.75	0.58
Personnel experience	1.35	1.19	1.00	0.87	0.71
Facilities	1.35	1.13	1.00	0.85	0.68
Schedule constraints	1.35	1.15	1.00	1.05	1.10
Degree of Planned					
Reuse	—	—	1.00	1.25	1.48
Teamwork	1.45	1.31	1.00	0.75	0.62
Process efficiency	1.35	1.20	1.00	0.85	0.65

Similar tables are available for the cost drivers in the COCOMO model but are embedded in the simulation tools, so the requirements engineer only has to select the ratings from the Likert scale.

In any case, the net result of a WEBMO calculation is a statement of effort and duration to complete the project in person-months and calendar months, respectively.

COSYSMO

COSYSMO (COnstructive SYstem engineering MOdel) is a new systems constructive cost model developed by Barry Boehm (inventor of COCOMO) and one of his students. COSYSMO is intended to be used for cost and effort estimation of software-intensive systems based on a set of size drivers, cost drivers, and team characteristics. The formulation of the COSYSMO metrics is similar to that for COCOMO, using Equations 10.1 and 10.2.

In COSYSMO, the size drivers include the counts of the following items as taken right from the SRS document:

- total system requirements
- interfaces
- operational scenarios
- the unique algorithms that are defined

Other drivers include

- requirements understanding
- architecture complexity
- level of service requirements
- migration complexity
- technology maturity

which are ranked using a Likert scale.

Finally, cost drivers based on team characteristics include

- stakeholder team cohesion
- personnel capability
- personnel experience/continuity
- process maturity
- multisite coordination
- formality of deliverables
- tool support

which are also rated on a likert scale.

COSYSMO is quite new, but it is already being used in a number of major corporations (Boehm et al. 2003).

Estimating Using Function Points

Function points were introduced in the late 1970s as an alternative to metrics based on source line count. This aspect makes function points especially useful to the requirements engineer. The basis of function points is that as more powerful programming languages were developed, the number of source lines necessary to perform a given function decreased. Paradoxically, however, the cost/LOC measure indicated a reduction in productivity, as the fixed costs of software production were largely unchanged (Albrecht 1979).

The solution to this effort estimation paradox is to measure the functionality of software via the projected number of interfaces between modules and subsystems in programs or systems. A big advantage of the function point metric is that it can be calculated during the requirements engineering activities.

Function Point Cost Drivers

The following five software characteristics for each module, subsystem, or system represent the function points or cost drivers:

- Number of inputs to the application (I)
- Number of outputs (O)
- Number of user inquiries (Q)
- Number of files used (F)
- Number of external interfaces (X)

In addition, the FP calculation takes into account weighting factors for each aspect that reflect their relative difficulty in implementation, and the function point metric consists of a linear combination of these factors, as shown in Equation 10.3.

$$FP = w_1 I + w_2 O + w_3 Q + w_4 F + w_5 X \tag{10.3}$$

where the w_i coefficients vary depending on the type of application system. Then complexity factor adjustments are applied for different types of application domains. The full set of coefficients and corresponding questions can be found by consulting an appropriate text on software metrics. The International Function Point Users Group maintains a Web database of weighting factors and function point values for a variety of application domains.

For the purposes of cost and schedule estimation and project management, function points can be mapped to the relative lines of source code in particular programming languages. A few examples are given in Table 10.2.

Now, these lines of code counts can be plugged into the COCOMO estimation equations to obtain appropriate estimates of effort for various features.

Here is an example of how this might work. A customer asks for a cost estimate for a desired feature. Based on the details of that feature, and the various weighting factors needed to calculate FP, the FP metric is computed. That number is converted to a lines of code count using the conversion shown in Table 10.2 (or another, appropriate conversion). These lines of code count number, along with the

Table 10.2 Programming Language and Lines of Code per Function Point

Language	*Lines of Code per Function Point*
C	128
C++	64
Java	64
SQL	12
Visual Basic	32

Adapted from Jones 1996.

various other aspects of the project, are plugged into a COCOMO estimator—which yields an estimate of time and effort (meaning person-hours) to complete the project. Of course, you wouldn't take this estimate for granted. It would be appropriate to check the estimate using other techniques that are complementary to the FP/COCOMO estimation approach (but this subject is out of scope). In any case, let's assume that you believe the estimate that COCOMO gives you. This estimate of number of person-months can be converted into an appropriate cost estimate for the customer, a task that would probably be done through the sales department.

Feature Points

Feature points are an extension of function points developed by Software Productivity Research, Inc. in 1986. Feature points address the fact that the classical function point metric was developed for management information systems and therefore are not particularly applicable to many other systems, such as real-time, embedded, communications, and process control software. The motivation is that these systems exhibit high levels of algorithmic complexity, but sparse inputs and outputs.

The feature point metric is computed in a similar manner to the function point except that a new factor for the number of algorithms, A, is added, yielding Equation 10.4.

$$FP' = w_1 I + w_2 O + w_3 Q + w_4 F + w_5 X + w_6 A \tag{10.4}$$

Use Case Points

Use case points (UCP) allow the estimation of an application's size and effort from its use cases. This is a particularly useful estimation technique during the requirements engineering activity when use cases are the basis for requirements elicitation.

The use case point equation is based on the product of four variables that are derived from the number of actors and scenarios in the use cases and various technical and environmental factors. The four variables are

1. Technical Complexity Factor (TCF)
2. Environment Complexity Factor (ECF)
3. Unadjusted Use Case Points (UUCP)
4. Productivity Factor (PF)

Leading to the use case point equation:

$$\text{UCP} = \text{TCF} \cdot \text{ECF} \cdot \text{UUCP} \cdot \text{PF} \tag{10.5}$$

This metric is then used to provide estimates of project duration and staffing from data collected from other projects. Use case points are a relatively new estimation technique and at this time, most information about this technique and a variety of free tools can be found on the Web.

Requirements Feature Cost Justification

Consider the following situation. A customer has the option of incorporating feature A into the system for $250,000 or forgoing the feature altogether. Currently $1,000,000 is budgeted for activity associated with feature A. It has been projected that incorporating the feature into the software would provide $500,000 in immediate cost savings by automating several aspects of the activity.

Should the manager decide to forego including feature A into the system, new workers would need to be hired and trained before they can take up the activities associated with feature A.* At the end of two years, it is expected that the new workers will be responsible for $750,000 cost savings. The value justification question is "should the new feature be included in the system or not?" We can answer this question after first discussing several mechanisms for calculating the value of certain types of activities, including the realization of desired features in the system.

Return on Investment

Return on investment (ROI) is a rather overloaded term that means different things to different people. In some cases it means the value of the activity at the time it is undertaken. To others it is the value of the activity at a later date. In other cases it is just a catchword for the difference between the cost of the system and the savings anticipated from the utility of that system. To still others there is a more complex meaning.

One traditional measure of ROI for any activity or system is given as

$$\text{ROI} = \text{Average Net Benefits}/\text{Initial Costs} \tag{10.6}$$

The challenge with this model for ROI is the accurate representation of average net benefit and initial costs. But this is an issue of cost accounting that presents itself in all calculations of costs versus benefits.

* Such a cost is called a "sunken cost" because the money is gone whether one decides to proceed with the activity or not.

Other commonly used models for valuation of some activity or investment include net present value (NPV), internal rate of return (IRR), profitability index (PI), and payback. We'll look at each of these shortly.

Other methods include Six Sigma and proprietary Balanced Scorecard models. These kinds of approaches seek to recognize that financial measures are not necessarily the most important component of performance. Further considerations for valuing software solutions might include customer satisfaction, employee satisfaction, and so on, which are not usually modeled with traditional financial valuation instruments.

There are other, more complex accounting oriented methods for valuing software. Discussion of these techniques is beyond the scope of the text. The references at the end of the chapter can be consulted for additional information (for example, Raffo et al. 1999; Morgan 2005).

Net Present Value

Net Present Value (NPV) is a commonly used approach to determine the cost of software projects or activities. The NPV of any activity or thing is based on the notion that a dollar today is worth more than that same dollar tomorrow. You can see the effect as you look farther into the future, or back to the past (some of you may remember when gasoline was less than $1 per gallon). The effect is due to the fact that, except in extraordinary circumstances, the cost of things always escalates over time.

Here is how to compute NPV. Suppose that FV is some future anticipated payoff either in cash or in anticipated savings. Suppose r is the discount rate* and Y is the number of years that the cash or savings is expected to be realized. Then the net present value of that payoff is

$$\text{NPV} = FV/(1 + r)^Y \tag{10.7}$$

NPV is an indirect measure because you are required to specify the market opportunity cost (discount rate) of the capital involved.

To see how you can use this notion in value engineering of requirements, suppose that you expect a certain feature, B, to be included in the system at a cost of $60,000. You believe that benefits of this feature are expected to yield $100,000 two years in the future. If the discount rate is 3%, should the feature be included in the new system?

To answer this question, we calculate the NPV of the feature using Equation 10.7, taking into account the cost of the feature

* The interest rate charged by the U.S. Federal Reserve. The cost of borrowing any capital will be higher than this base rate.

$$\text{NPV} = 100{,}000/1.03^2 - 60{,}000 = 34{,}259$$

Since NPV is positive, yes, the feature should be included in the requirements.

Equation 10.7 is useful when the benefit of the feature is realizable after some discrete period of time. But what happens if there is some incremental benefit of adding a feature; for example, after the first year, the benefit of the feature doubles (in future dollars). In this case we need to use a variation of Equation 10.7 that incorporates the notion of continuing cash flows.

For a sequence of cash flows, CF_n, where $n = 0,\ldots,k$ represents the number of years from initial investment, the net present value of that sequence is

$$\text{NPV} = \sum_{n=0}^{k} \frac{CF_n}{(1+r)^n} \tag{10.8}$$

The CF_n could represent, for example, a sequence of related expenditures over a period of time, such as features that are added incrementally to the system or evolutionary versions of a system.

Internal Rate of Return

The internal rate of return (IRR) is defined as the discount rate in the NPV equation that causes the calculated NPV, minus its cost, to be zero. This can be done by taking Equation 10.7 and rearranging to obtain

$$0 = FV/(1 + r)^Y - c$$

where c is the cost of the feature and FV its future value. Solving for r yields

$$r = [FV/c]^{1/Y} - 1 \tag{10.9}$$

For example, to decide if we should incorporate a feature or not, we compare the computed IRR of including the feature with the financial return of some alternative, for example, to undertake a different corporate initiative or add a different feature. If the IRR is very low for adding the new feature, then we might simply want to take this money and find an equivalent investment with lower risk. But if the IRR is sufficiently high for the new feature, then the decision might be worth whatever risk is involved in its implementation.

To illustrate, suppose that a certain feature is expected to cost \$50,000. The returns of this feature are expected to be \$100,000 of increased output two years in the future. We would like to know the IRR on adding this feature.

Here, NPV = 100,000/(1 + r)^2 – 50,000. We now wish to find the r that makes the NPV = 0, that is, the "break even" value. Using Equation 10.9 yields

$$r = [100{,}000/50{,}000)]^{1/2} - 1$$

This means r = 0.414 = 41.4%. This rate of return is very high for this feature, and we would likely choose to incorporate it into the system.

Profitability Index

The profitability index (PI) is the NPV divided by the cost of the investment, I:

$$\text{PI} = \text{NPV/I} \quad (10.10)$$

PI is a "bang-for-the-buck" measure, and it is appealing to managers who must decide between many competing investments with positive NPV financial constraints. The idea is to take the investment options with the highest PI first until the investment budget runs out. This approach is not bad but can sub-optimize the investment portfolio.

One of the drawbacks of the profitability index as a metric for making resource allocation decisions is that it can lead to a sub-optimization of the result. To see this effect, consider the following situation. A customer is faced with some hard budget-cutting decisions, and she needs to remove some features from the proposed system. To reach this decision she prepares the analysis based on the profitability index shown in Table 10.3.

Suppose the capital budget is $500K. The PI ranking technique would cause the customer to pick A and B first, leaving inadequate resources for C. Therefore, D will be chosen, leaving the overall NPV at $610K. However, using an integer programming approach will lead to a better decision (based on maximum NPV) in that features A and C would be selected, yielding an expected NPV of $620K.

Table 10.3 A Collection of Prospective Software Features, Their Project Cost, NPV, and Profitability Index

Feature	*Projected Cost (in $10s of thousands)*	*NPV (in $10s of thousands)*	*PI*
A	200	260	1.3
B	100	130	1.3
C	300	360	1.20
D	200	220	1.1

The profitability index method is helpful in conjunction with NPV to help optimize the allocation of investment dollars across a series of feature options.

Payback Period

To the project manager, the payback period is the time it takes to get the initial investment back out of the project. Projects with short paybacks are preferred, although the term "short" is completely arbitrary. The intuitive appeal is reasonably clear: the payback period is easy to calculate, communicate, and understand.

Payback can be used as a metric to decide whether to incorporate a requirement in the system or not. For example, suppose implementing feature D is expected to cost $100,000 and result in a cost savings of $50,000 per year. Then the payback period for the feature would be two years.

Because of its simplicity, payback is the least likely ROI calculation to confuse managers. However, if payback period is the only criterion used, then there is no recognition of any cash flows, small or large, to arrive after the cutoff period. Furthermore, there is no recognition of the opportunity cost of tying up funds. Since discussions of payback tend to coincide with discussions of risk, a short payback period usually means a lower risk. However, all criteria used in the determination of payback are arbitrary. And from an accounting and practical standpoint, the discounted payback is the metric that is preferred.

Discounted Payback Period

The discounted payback period is the payback period determined on discounted cash flows rather than undiscounted cash flows. This method takes into account the time (and risk) value of the sunken cost. Effectively, it answers the questions, "how long does it take to recover the investment?" and "what is the minimum required return?"

If the discounted payback period is finite in length, it means that the investment plus its capital costs are recovered eventually, which means that the NPV is at least as great as zero. Consequently, a criterion that says to go ahead with the project if it has *any* finite discounted payback period is consistent with the NPV rule.

To illustrate, in the payback example just given, there is a cost of $100,000 and an annual maintenance savings of $50,000. Assuming a discount rate of 3 percent, the discounted payback period would be longer than two years because the savings in year two would have an NPV of less than $50,000 (figure out the exact payback period for fun). But because we know that there is a finite discounted payback period, we know that we should go ahead and include feature D.

Exercises

10.1 Why is it so important to determine the cost of features early, but not too early in the requirements engineering process?

10.2 What factors determine which metric or metrics a customer can use to help make meaningful cost–benefit decisions of proposed features for a system to be built?

10.3 How does the role of ranking requirements help in feature selection cost–benefit decision making?

10.4 What changes (if any) would you need to make to the COCOMO or feature point equation calculations to incorporate ranking of requirements?

10.5 Investigate the use of other decision-making techniques, such as integer programming, in helping to decide on the appropriate feature set for a proposed system.

10.6 Complete the derivation of Equation 10.8 from Equation 10.6 by setting the NPV equation (less the cost of the investment) to zero and solving for *r*.

References

Albrecht, J. (1979) Measuring application development productivity, in *Proc. IBM Applications Develop. Symp.*, Monterey, CA, Oct. pp. 14–17.

Boehm, B.W., D.J. Reifer, and R. Valerdi (2003) COSYSMO: A Systems Engineering Cost Model, *Proceedings of the First Conference on Systems Integration.* On line at http://valerdi.com/cosysmo/, last accessed 1 September 2008.

Jones, C. (1996) Activity-based software costing. *IEEE Computer*, May, pp. 103–104.

Laplante, P.A. (2006) *What Every Engineer Needs to Know About Software Engineering*, CRC/Taylor & Francis.

Laplante, P.A. (2003) *Software Engineering for Image Processing Systems*, CRC Press.

Morgan, J.N. (2005) A roadmap of financial measures for IT project ROI. *IT Professional*, Jan./Feb., pp. 52–57.

Raffo, D., J. Settle, and W. Harrison (1999) *Investigating Financial Measures for Planning of Software IV&V*, Portland State University Research Report #TR-99-05.

Reifer, D.J. (2002) Estimating web development costs: There are differences. On line at http://www.reifer.com/download.html.

Appendix

Software Requirements Specification for a Smart Home

Version 2.0

September 20, 2008

1 Introduction

1.1 Purpose: Mission Statement

Making residential enhancements that will pave the way for an easy and relaxed transition into retired life.

Document prepared for the Smith family home, a pre-existing building.

1.2 Scope

The "Smart Home" system, herein referred to as "The System," will be a combination of hardware and software that will provide an escape from daily routines and mundane tasks. This product seeks out the items that consume the most time but do not need to. The goal is to automate that which does not really need human interaction, to free the occupants to enjoy themselves in their retirement. The system will not free the mundane household chores from any human interaction, but it will require only as little as needed.

1.3 Definitions, Acronyms, and Abbreviations

pH—see http://en.wikipedia.org/wiki/PH
RFID—Radio Frequency Identification
SH—Smart Home
SAN—Storage Area Network
SRS—System Requirements Specification
WPA—Wi-Fi Protected Access
WEP—Wired Equivalent Privacy
USB—Universal Serial Bus

1.4 References

802.11 IEEE Specification http://ieeexplore.ieee.org/servlet/opac?punumber=4248376

1.5 Overview

Requirements have been divided into key functional areas, which are decomposed into features within those functional areas. Functional and nonfunctional requirements exist within the laid-out document format. The order of the leading sections and the corresponding requirements should be interpreted as priority.

2 Overall Description

2.1 Product Perspective

This system consists of many standalone devices. At this time it is not known if these devices exist in the commercial market or not. The document has a holistic approach, intermingling the demand for devices with the functions of the system. The first step in continuing with this project would be to decide on feasibility and do some cost analysis for some of the requirements contained herein.

This document seeks to lay out areas where all interfaces are abstracted. There are areas where there would clearly need to be communication between the various interfaces, but since there is no targeted device, there is no known protocol for speaking with that device.

2.2 Product Functions

The functions of this product will be divided into six categories. Accessibility is the first and most highly desired by the customer. This functional category seeks to improve the user experience of the system by providing various benchmarks for deciding on usability. The second major functional area is environmental considerations. The aim for this area is to ensure the inhabitants have a safe environment in which to live, and the SH system enhances this environment instead of adding risks or hazards to the environment. The most important aspects for this category within this document will be monitoring and helping quality of air and water. The third category is energy efficiency. It is desired by the system and the customers to not only enhance their lives while living in this SH, but also to live in an efficient way. This system will not only monitor the occupants' energy usage, but it will also seek to improve the occupants' abilities to save on energy costs. Fourth we have security. Security is important to the customers as they want their home to be safe. The security system in the SH will provide added layers of protection from various crimes as they are happening, but also add layers to help prevent crimes from happening in the first place. The security will also give the occupants more peace of mind as they will have far greater control and oversight should they ever need to go for some extended trip away from their SH. The fifth section deals with media and entertainment. The goal for this section of the system is to decentralize home entertainment and make it available to the occupants wherever they desire. Finally, there will be automation. This is the section that will get into the guts of what is meant by taking as much of the human element out of routine tasks as possible. The summation and harmonization of all the six categories of the SH will provide for a truly rewarding living experience for the occupants of the SH.

2.3 User Characteristics

The primary users of this system will be two older adults entering retirement. One of the adults spent his work life doing IT support and has a mild degree of electronic and computer expertise. The other was a school teacher and is not very familiar or comfortable with electronic and computing devices. These individuals are both of sound physical abilities, although one is a little shorter and suffers from sporadic hip pains.

2.3.1 User/Stakeholder Profiles

Stakeholder	*Interests*	*Constraints*
Local building codes	Ensuring the safety for the building for the inhabitants.	Multiple business codes, especially around electrical interfaces.
Robert and Elizabeth Smith	Inhabitants interested in easing their lives.	None.

(continued on next page)

Stakeholder	*Interests*	*Constraints*
Interior designer	Ensuring the functionality of the system does not deter from the esthetic.	None.
Building architect	Ensuring the existing structure can support the improvements.	None.
Construction workers	Making sure the construction details are clearly identified.	None
Developers	Making sure interfaces are defined.	None.
Tim Smith [son]	Ease of use for occasional user.	None.
Cats	Safety.	None.
Relatives	Ease of use and comfort.	One relative in wheel chair.
House sitters	Easily understanding limited sets of functionality.	None.
Guests	Ease of use and comfort.	None.
Maid service	Minimal functional understanding.	None.
Utility company	Negative for alternative energy, sharing alternate use policies.	None.
Internet services provider	Making sure bandwidth and services are available.	None.
Tivo	Negative, will lose business.	None.

2.4 Constraints

IEC 61508—Providing for functional safety.

None others at this time, as feasibility and cost estimation activities are out of scope.

2.5 Assumptions and Dependencies

All hardware is available.
All devices will present the data listed below.
Occupants will provide feed elements for the devices that require them.

3 Core System requirements

This section will list all of the core functional requirements for the SH system.

3.1 Central Processing Requirements

3.1.1 System shall operate on a system capable of multi-processing.

3.1.2 System shall operate on a system capable of near real time execution of instructions.

- 3.1.2.1 System shall service triggers or stimuli in no more than 500 milliseconds.

3.1.3 System shall operate in a highly available and fault tolerant manner.

- 3.1.3.1 System shall have a reported uptime of 99.9% (4 NINES).
- 3.1.3.2 System shall recover from locked state in less than one (1) second.

3.1.4 System shall have a database associated with it that can handle transaction processing at a rate of one thousand transactions per minute.

3.1.5 System shall have redundant databases for fail over purposes.

3.1.6 System shall perform periodical offsite and onsite backups of all configuration and reporting data.

3.1.7 System shall support wireless encryption protocols WPA [1-2] and WEP.

3.1.8 System shall support wired Ethernet for 1 gigabit per second, and contain cat 6e cabling.

3.1.9 System may contain separate SAN device for storage flexibility.

3.1.10 System may contain separate video recorder/processor for process distribution.

3.1.11 If system supports recording more than three (3) television shows simultaneously, then system shall have separate video recorder.

4 Accessibility

Accessibility is defined as the need of the SH system to be usable by all persons, including those with any physical impairments or those with difficulty operating and/or understanding complex electronic systems. Priority = High.

4.1 Use of SH Features

4.1.1 SH System shall be usable by those with slight eye loss.
- 4.1.1.1 System shall not have any buttons smaller than one (1) inch square.
- 4.1.1.2 System shall have all consoles and controlling devices between four (4) and five (5) feet from ground level.
- 4.1.1.3 System shall have backlighting on all buttons for nighttime ease of use.
- 4.1.1.4 System shall have options to increase and decrease font sizes on Web interfaces and all console and controlling devices.
- 4.1.1.5 System shall have liquid layouts for all graphical interfaces for display on many different types of display devices.

4.1.2 System shall be easy to use.
- 4.1.2.1 System shall be understood by users of all levels of understanding with no more than four (4) hours of training.
- 4.1.2.2 System shall have a help function associated with all user entry possibilities.
- 4.1.2.3 System shall have text to speech capabilities to allow the user to receive vocal instructions for help and how to menu items.

5 Environment

Environment encompasses air quality controls, but also includes other environmental elements such as lighting levels, water purification, etc. Priority = High.

5.1 Water, Air Purification

Water purification and air quality are key factors to a good environment within the SH. This system seeks to improve by monitoring both air quality and water purification. Priority = High.

5.1.1 SH shall have a reverse osmosis water purification system.
5.1.2 SH shall have a nonfiltered water system.
5.1.3 System shall store how much water passes through the filtration unit each day.
5.1.4 System may send notifications to users about how much water passes through the filtration system.
5.1.5 System shall have the option to send notifications to users when water filtration unit needs cleaning.

5.1.6 System shall incorporate water softener system into water system.
5.1.7 System shall monitor the salt in the water softener.
5.1.8 System shall accept user input for desirable levels of salt in the water softener device.
5.1.9 System shall send notifications to users when salt in softener gets below user defined levels.
5.1.10 System shall monitor air filter.
5.1.11 System shall send notification to users when air filter needs to be cleaned and/or changed.
5.1.12 System shall provide monitors for measuring air quality.
5.1.13 System shall accept user input for air quality thresholds.
5.1.14 System shall notify users when air quality reaches levels outside the user defined thresholds.

5.2 Health and Safety Detectors

This section describes the role and interfaces for various common household detectors. The detectors carry out their common functionality, but reactions are automated, and historical data are logged. Priority = High.

5.2.1 SH shall have at least one (1) multi purpose detector for detecting smoke and carbon monoxide on each floor.
5.2.2 System shall not interfere in any way with detector's manufacturer's operating procedures.
5.2.3 System shall accept user input for dangerous levels of smoke and carbon dioxide.
5.2.4 System shall trigger warning and require additional confirmation when users select levels outside of the manufacturer's settings for dangerous levels of smoke and carbon dioxide.
5.2.5 System shall notify proper authorities when levels above user defined thresholds of smoke or carbon monoxide are detected.
5.2.6 System shall tie into detectors to send remote alert messages to users when elevated levels of smoke or carbon monoxide are detected.
5.2.7 System shall utilize radon detector in the basement.
5.2.8 System shall allow users to set defined ceiling for radon levels.
5.2.9 System shall accept input from users for notification events for radon level detections.
5.2.10 System shall send notifications based on user defined notification events to interested parties when radon levels are more than the user defined ceiling.
5.2.11 System shall activate basement fan system when radon levels report above defined ceiling.
5.2.12 System shall record radon levels routinely.

5.2.13 System shall allow users to view reports on radon levels.
5.2.14 System shall persist radon level data for no less than ninety (90) days.

6 Energy Efficiency

Energy efficiency covers the extent to which the SH system enables the users to monitor and enhance energy efficiency of the house through "smart" and adaptive controls and interfaces. Priority = High.

6.1 Air Conditioner/Heating

Controlling and adapting the air conditioning and heating are important aspects of using energy efficiently. Not only does the SH seek to improve the ease of use of traditional thermostats, but also to provide intelligence in order to optimize the use of the system. Priority = High.

6.1.1 SH shall be divided up into zones for heating and cooling.
6.1.2 System shall accept desired temperature settings for each zone, for no less than four (4) periods in the day.
6.1.3 System shall accept input for desired room temperature when room is unoccupied.
6.1.4 System shall detect motion to determine if a room is occupied, and make proper adjustments to the temperature.
6.1.5 System shall differentiate between pets and occupants for motion detection and temperature adjustment.
6.1.6 System shall monitor outdoor temperature and humidity.
6.1.7 System shall shut down air conditioning and open windows if temperature outside is cooler than the inside temperature.
6.1.8 System shall not open or close any windows if there is something in the desired path of the window.
6.1.9 System shall reverse directions of windows if they encounter any resistance.
6.1.10 System shall send notifications to users if windows need to reverse path or windows cannot complete desired action (open or close).

6.2 Time-of-Day Usage

Time-of-day usage refers to common utility programs that provide reduced rates for utilities used during off-peak time periods.

6.2.1 Appliances shall be configured if they need to be used as time of day devices.
6.2.2 System shall accept definitions for the range for time of day savings.

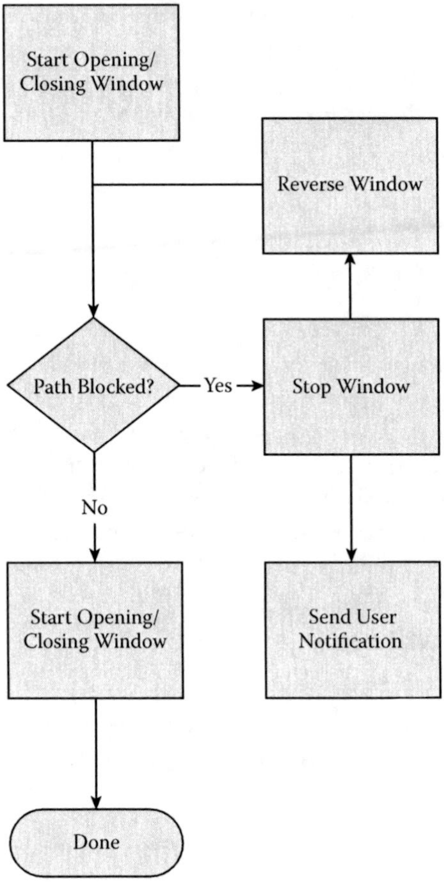

Figure A.1 Window movement flow chart.

6.2.3 System shall queue up appliance(s) to run when time of day period starts if it is a time of day device.
6.2.4 System shall allow user to override the time of day setting to run the device/appliance immediately.
6.2.5 System may send notification when device has completed its work.

6.3 Water Recovery

Promote reusable resources by capturing rain water to use for irrigation. Priority = Medium.

6.3.1 System shall have water recovery system for rain water.
6.3.2 System shall use water recovered from rain for lawn irrigation.

6.3.3 System shall record amounts of rain water recovered on a monthly basis.
6.3.4 System shall allow user to view reports for rain water recovery.
6.3.5 System shall persist rain water recovery data for no less than twenty-four (24) months.

6.4 Alternative Energy

Interface to allow for future expansion and addition of alternative energy sources. Priority = Low.

6.4.1 System shall provide interface into central electrical supply for alternative energy to supply power to the house (i.e., solar or wind).
6.4.2 System shall monitor the generation of alternative energy.
6.4.3 System shall present reports to users of the amounts of alternative energy generated during some user-defined time period.
6.4.4 System shall maintain alternative energy generation data for no less than two (2) years.

6.5 Air Flow Monitoring

The air flow monitoring system serves as a detection agent for wasted energy. The monitoring of air flow, especially in conjunction with running central air conditioning or heating, will lead to discovery of drafts and leaks. Priority = Medium.

6.5.1 System shall monitor air flow in various rooms within the SH.
6.5.2 System shall present reports for air flow to users.
6.5.3 System shall persist air flow data for no less than three (3) years.
6.5.4 System shall accept input for thresholds for detecting drafts or leaks of air within the house.
6.5.5 Assuming the house is running centralized air or heat, system shall send notifications if drafts are detected that exceed the user-defined threshold.

7 Security

Security includes aspects of home security like alerts in the event of a break-in, video monitoring of premises or areas of interest, as well as unattended monitoring while the occupant(s) is away. Priority = High.

7.1 Home Security

Home security centers around controlling the access points to the SH as well as providing many cameras to provide views into areas of the SH. The SH will provide automated and human commanded responses to various security situations. Priority = High.

7.1.1 System shall have biometric and keypad door locks for all points of entry into the house.
7.1.2 System shall allow users to encode a one time use door code for expected visitors.
  7.1.2.1 System shall allow users to remotely code a one time use code for visitors (i.e., over the phone, Internet, or some other mobile device).
7.1.3 System shall record all entries based on code entered or biometrics presented.
7.1.4 System shall present report to users for all entries.
7.1.5 System shall persist home entry data for no less than ten (10) years.
7.1.6 System shall allow for RFID tags to open garage doors.
7.1.7 System shall allow for biometric and key pad entry to open garage doors.
7.1.8 System shall allow user to configure maximum duration for garage door to remain open.
7.1.9 System shall shut open garage door if it is open past the user-defined maximum.
7.1.10 System shall allow user to override automatically shutting the garage door, i.e., "Hold Open option".
7.1.11 Garage door shall reverse course if something is blocking its path.
7.1.12 System shall notify user if the garage door is unable to safely close.
7.1.13 System shall allow users to configure entry routine for all RFID, biometric, and key codes, i.e., upon entry for user X through the garage door, turn on garage light, hall light, and kitchen light.

7.2 Unattended Home

Unattended Home is a set of responses to various triggers throughout the home and immediate responses to those triggers. This will aid in security of the home. Priority = High.

7.2.1 System shall allow users to set an away mode.
7.2.2 System shall define away mode as time and date range when users will be away from their house.
7.2.3 System shall allow users to configure lights in any room to go on for some defined duration while they are away.
7.2.4 System shall deploy motion detectors to always be on for the duration while the user is away.
7.2.5 System shall differentiate between motion detected for pets and humans.

7.2.6 System shall send notification to users if any motion detectors are triggered while the user is away.
7.2.7 User shall be presented options to view various cameras via the Web or some other mobile device when motion detectors are triggered.
7.2.8 User shall be given the option to alert the authorities when motion detectors are triggered.
7.2.9 System shall turn on user-defined lights in and outside the house when motion detectors are triggered.

7.3 Monitoring Anywhere

Occupants and users of the SH's system should be able to monitor the home from anywhere they wish. This includes many different cameras within the SH as well as various points of entry and other triggers placed throughout the home. This will give the occupants more freedom to travel while feeling their home is secured and well cared for. Priority = Medium.

7.3.1 System shall show camera data streams to any television in the house.
7.3.2 System shall incorporate cameras at points of entry with doorbells to allow users to view visitor.
7.3.3 System shall allow user to remotely unlock the door to permit entry to the visitors.
7.3.4 System shall allow users to notify emergency personnel of possible intruder.
7.3.5 System shall permit users to view security cameras from a secure Web site or mobile device for remote viewing of property.

8 Media/Entertainment

Media and entertainment include the ability to create, store, and access multiple forms of media and entertainment such as audio, video, etc. anywhere in the house. Priority = Medium.

8.1 Recording Television Shows

Recording television shows allows the users to throw out the VCR and gives them a more automated and intelligent solution for recording all their favorite shows or movies that play through the television. Priority = Medium.

8.1.1 System shall allow user to record any show on television.
8.1.2 System shall present a Web interface with a grid listing similar to the TV Guide book for users to select shows to records.

8.1.3 System shall allow user to record a minimum of two (2) television shows simultaneously.
8.1.4 System shall make storage for recorded shows expandable.
8.1.5 System shall free storage space as needed by first-in first-out (FIFO) or some other defined priority schedule.
8.1.6 System shall provide search feature to search through television shows to select which one to record.
8.1.7 System shall provide user the ability to record all occurrences of a specified show.
8.1.8 System shall provide user the ability to record only new instances of a specified show.
8.1.9 System shall provide telephone menu options for customer to dial in and select channel, time, and duration to record.
8.1.10 System shall present users option to select quality for recording.
8.1.11 System shall present user option to not automatically overwrite the television recording.
8.1.12 System shall give user the option to only store X number of episodes from a certain series at a time.
8.1.13 System may skip commercials when system is able to detect the commercial.
8.1.14 System shall monitor storage space for future recordings.
8.1.15 System shall send notification when resources get low enough where recordings will be overwritten.
8.1.16 System shall permit users to not automatically delete a show or a series.
8.1.17 System shall not record any new shows if there is space available for recovery.
8.1.18 System shall send notifications to users if there is no longer space available to record new shows.

8.2 Video Entry

Video Entry is the mechanism by which various formats of video data are able to be loaded into the repository for video playback. Priority = Medium.

8.2.1 System shall allow for video input into digital library.
8.2.2 System shall allow for storage of video meta data such as category, genre, title, rating, etc.
8.2.3 System shall provide interface to users to edit and update video meta data.
8.2.4 System shall accept one-button touch support for incorporating VHS tape into digital library.
8.2.5 System shall accept one-button touch support for incorporating DVD videos into digital library, where law and technology provide.

8.3 Video Playback

Video playback allows the users of the SH to be able to enjoy video, both recorded and preloaded content from anywhere within the house. Priority = Medium.

- 8.3.1 System shall allow for recorded video playback at any television in the house.
- 8.3.2 System shall allow for other video media to be available for playback in any room with a television.
- 8.3.3 System shall allow of all common features of a VCR player or DVD player, such as fast forward, rewind, chapter skip, etc.
- 8.3.4 System shall allow user to skip commercials where commercials are detected.
- 8.3.5 System shall prevent user from playing back identical media on multiple televisions at the same time.
- 8.3.6 System shall allow user to remove the recording from storage when they are done watching.
- 8.3.7 System shall allow user to remove other video media from the storage.

8.4 Audio Storage and Playback

The audio storage and playback are important features of the smart home. This section will describe how the audio is imported into the digital library as well as what capabilities there are for distributing or sharing the audio either to various rooms in the SH or to external media. Priority = Medium.

- 8.4.1 System shall accept input into digital audio library from CD.
  - 8.4.1.1 System shall allow user to enter a CD into a tray and immediately rip the CD.
  - 8.4.1.2 System shall collect all available meta data from the CD from the Internet for categorization.
  - 8.4.1.3 System shall store audio binary in a lossless format.
- 8.4.2 System shall accept input into the digital audio library from a USB device.
- 8.4.3 System shall provide interface for users to manually place audio file into digital audio library.
- 8.4.4 System shall automatically normalize volume of all audio files loaded into the digital audio library.
- 8.4.5 System shall store information about audio files in some searchable entity.
- 8.4.6 System shall provide users the ability to alter meta data for any file in the collection.

8.4.7 System shall allow users to remove audio files from the digital audio library.

8.4.8 System shall allow for categorization of audio files by important fields such as genre, artist, album, etc.

8.4.9 System shall allow audio playback.

- 8.4.9.1 System shall allow for wired or wireless connection to any device capable of audio playback.
- 8.4.9.2 System shall allow centralized panel to play back various audio files in different rooms of the house.
- 8.4.9.3 System shall provide access point in garage so digital audio can be downloaded to an automobile audio system.

8.4.10 System shall allow users to author new CDs.

- 8.4.10.1 System shall allow users to select tracks for newly authored CD from a playlist or from the complete library.
- 8.4.10.2 System shall allow user to select which format to use.
- 8.4.10.3 System shall allow user to select a CD burning drive.
- 8.4.10.4 System shall provide guidance to users for available space depending on the drive and media selected as well as the format chosen.
- 8.4.10.5 System shall verify proper media is in the selected drive.
- 8.4.10.6 System shall allow user to select order for the tracks.
- 8.4.10.7 System shall allow user to confirm track information to start authoring the CD.
- 8.4.10.8 System shall perform necessary audio conversion and burn the CD based on the user's authoring details.
- 8.4.10.9 System may notify user when the authoring process is complete.

8.4.11 System shall allow users to create, edit, and delete playlists.

- 8.4.11.1 Playlist shall consist of one to "N" number of tracks selected from the digital library.
- 8.4.11.2 A single track may reside in any number of playlists.
- 8.4.11.3 A single track may not reside in an individual playlist more than once.
- 8.4.11.4 System shall allow users to set a name and description for all playlists created.

8.4.12 System shall allow users to transfer music from the digital library to portable music players.

- 8.4.12.1 System shall allow track transfer according to both complete playlists as well as individual tracks.
- 8.4.12.2 System shall allow users to format or delete the current selection of tracks on the portable device before transferring.

8.4.12.3 System shall allow users to append additional tracks or playlists to the portable device as space permits.

8.4.12.4 System shall add tracks to the device in the order they were selected by the user until all tracks are transferred or the device is full.

9 Automation

Automation is the processes of making something routine happen automatically, on some set schedule or by some defined trigger such that little or no human intervention is needed. Priority = Medium.

9.1 Pet Care

The pet care system will be added to ensure that proper unattended care of the pets in the household takes place. In this instance cats are the pets within the household, but the system should extend to other future pets as well. The goal is to have a mostly automated system to take care of the pets' primary needs such as food, water, and waste disposal, but the system can also track health care needs such as appointments and vaccinations. Priority = High.

9.1.1 System shall handle providing water for the pets.

9.1.1.1 Pet watering bowls shall be tied into the water filtration system [ref. requirement 5.1.1].

9.1.1.2 System shall monitor consumption of water on a daily basis for pet bowls.

9.1.1.3 System may send time-defined notifications to users detailing the water consumption by pets.

9.1.1.4 System shall present a report for water consumption by pets.

9.1.1.5 Pet water consumption data shall persist for no less than thirty (30) days.

9.1.2 System shall provide food for pets.

9.1.2.1 System shall accept user input for intervals to deliver food.

9.1.2.2 System shall notify users when food in storage reaches low levels, as users will be required to fill the storage depot.

9.1.2.3 Pet's food shall be delivered to their bowls every user-defined interval.

9.1.2.4 System shall allow user to set portion weight for every bowl for the system.

9.1.2.5 Pet food delivery shall be portioned to user-defined weight.

9.1.2.6 Pet food delivery shall not exceed the amount of the portion weight; weight includes food already in bowl.
9.1.2.7 Food consumption shall be recorded per pet every feeding cycle.
9.1.2.8 Alert messages shall be sent if food delivery system dispenses no food for three (3) consecutive cycles.
9.1.2.9 System shall present a report for food consumption per pet.

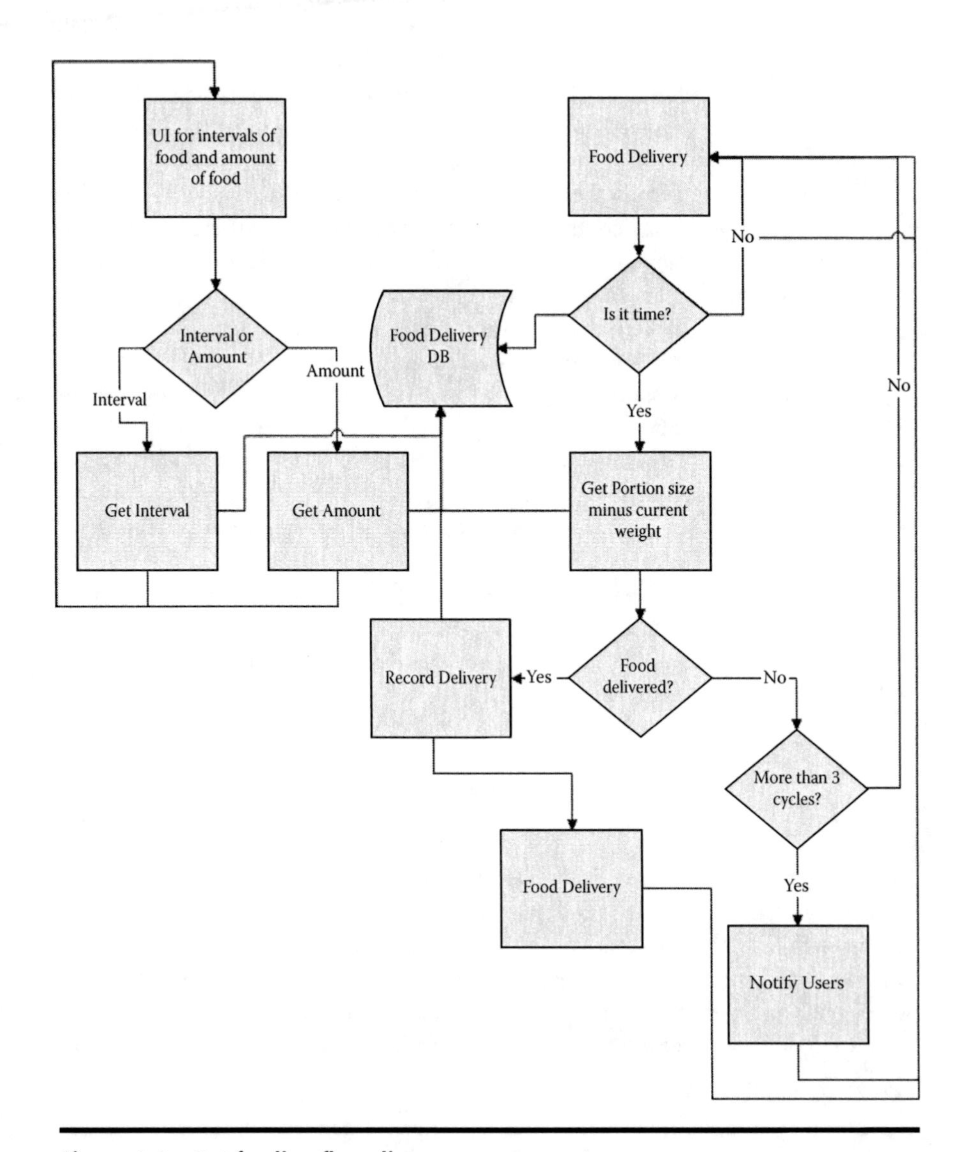

Figure A.2 Pet feeding flow diagram.

9.1.2.10 Pet food consumption data shall persist no less than thirty (30) days.
9.1.2.11 Pets shall wear RFID tags on their collars.
9.1.2.12 Pet food bowls shall open only when proper RFID tag is present.

9.1.3 System shall monitor and maintain pet litter box(es).
9.1.3.1 Pet litter box shall be cleaned when odor levels reach a user-defined mark, and litter disposal unit is not full.
9.1.3.2 System shall notify users if odor levels are above defined ceiling for more than eight (8) hours.
9.1.3.3 System shall notify users every two (2) hours when litter levels are low and continue to send alerts until the litter levels are within configurable ranges.
9.1.3.4 System shall notify users every four (4) hours when litter disposal container (where dirty litter is stored) is full, and continue to send alerts until the litter disposal unit is not full.

9.1.4 System shall monitor pet's health.
9.1.4.1 System may incorporate weight pad to measure weight while pet is feeding (pet based on RFID).
9.1.4.2 System may send user-defined notifications of weight change.
9.1.4.3 System may maintain weight data for no less than thirty (30) days.
9.1.4.4 System shall accept user input for types and intervals for vaccinations.
9.1.4.5 System shall accept input for when vaccinations have been administered.
9.1.4.6 System shall send notifications when vaccinations are more than one (1) week overdue.
9.1.4.7 System shall maintain vaccination records for no less than five (5) years.

9.2 Making Coffee

The coffee-making system will provide an automated process to make coffee for users. The system will not be completely autonomous as it does have external dependencies such as power, water supplies, and the users maintaining proper levels of coffee beans in the repository. The system will, however, expose many configuration options to begin the process as part of a timed sequence or through some other stimulus. Priority = Medium.

9.2.1 Coffee maker shall be tied into the water purification system.
9.2.2 System shall start coffee maker at any user-defined time as long as water is present, coffee bean levels are sufficient, and unit is powered.

9.2.3 System shall send a notification when bean levels are low.

9.2.4 When bean levels are too low to make coffee, system shall send an alert and coffee maker shall blink a warning indicator.

9.2.5 Coffee maker shall use a reusable filter.

9.2.6 System shall send notification when filter should be cleaned or changed.

9.2.7 Coffee maker shall shut off if weight measured by burner plate is less than the weight of the carafe or more than the weight of a full carafe.

9.2.8 Coffee maker shall have an emergency stop button clearly visible and accessible.

9.2.9 Coffee maker shall stop within one (1) microsecond when button is pressed.

9.2.10 Coffee maker shall turn off if carafe is not removed from burner plate for thirty (30) consecutive minutes.

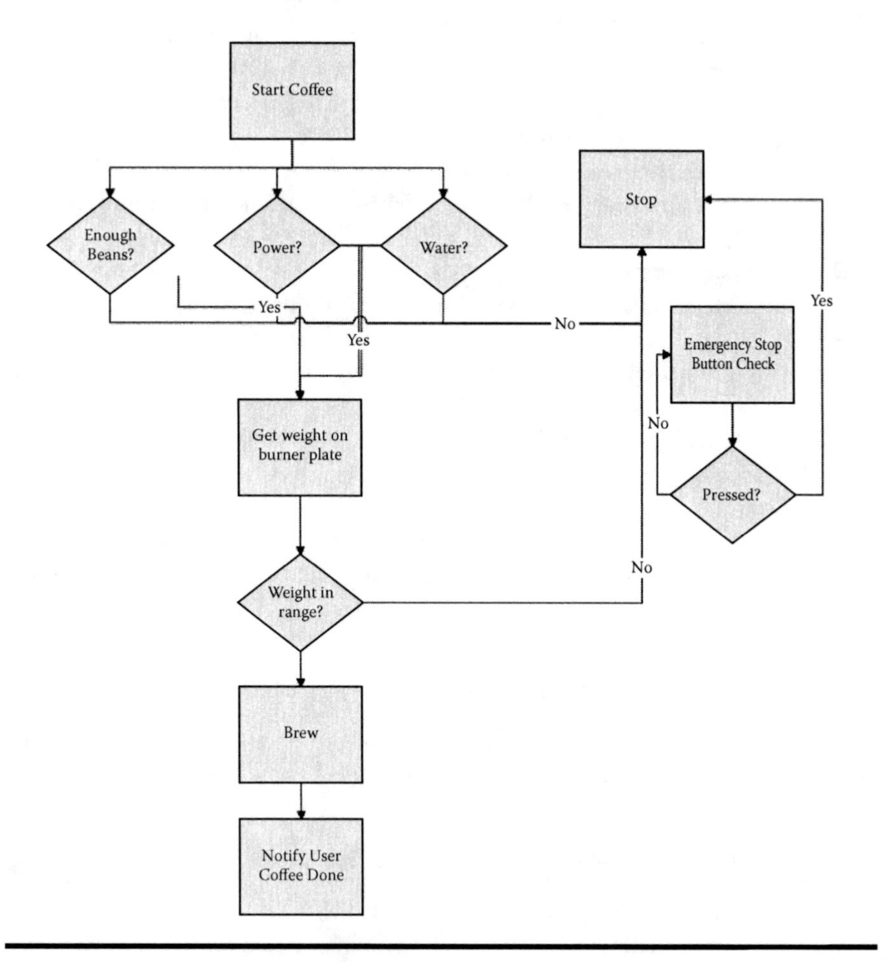

Figure A.3 Making coffee flow.

9.3 Hot Tub Maintenance

Hot tub maintenance is important to the users as this process will enable them to have little interaction in the daily needs of maintaining the hot tub. The process once again is not completely autonomous as it has external dependencies such as available water, power, and the users maintaining proper levels of chemicals needed for maintaining the hot tub. Priority = Medium.

9.3.1 System shall monitor temperature of hot tub water at all times.
9.3.2 Hot tub cover shall open with proper biometric credentials or proper entry of code on numeric key pad.
9.3.3 Hot tub cover shall not open if water temperature is not within a user-defined safe range.
9.3.4 System shall monitor pH and bacterial levels of hot tub water.
9.3.5 Hot tub cover shall not open if pH and bacterial levels are outside user-defined norms.
9.3.6 System shall administer chemicals to the hot tub water to maintain proper pH and bacterial levels.
9.3.7 System shall notify users when chemical levels are low.
9.3.8 System may send notifications of chemical administration information.
9.3.9 Hot tub shall be tied into central water system.
9.3.10 System shall monitor water levels of hot tub.
9.3.11 System shall replenish water to maintain proper water level.
9.3.12 System may send notifications of water level replenishment.
9.3.13 Hot tub cover shall not open if water level is outside user-defined norms.
9.3.14 Hot tub cover shall close with button press or if no activity/motion is detected for some time range, and water displacement levels are normal (no one in the tub).
9.3.15 Hot tub shall sound alarm if no activity or motion is detected for some defined time range and water displacement indicates there is someone in the tub.

9.4 Home Indoor Irrigation

The home irrigation system aims to provide care for household plants in an unattended fashion. A source of water will be an external dependency for this system. Priority = Low.

9.4.1 System shall control any number of indoor irrigation access points.
9.4.2 All access points shall be accompanied with soil moisture detector.
9.4.3 System shall allow user to define desired moisture level and watering intervals for each watering access point in the house.

9.4.4 System shall control watering units at each access point to keep soil at a steady moisture level.
9.4.5 If plants are on an interval, system shall bring the moisture level up to defined range during each watering interval.
9.4.6 System shall record water used and average soil moisture levels per access point.
9.4.7 System shall maintain indoor watering information for no less than forty-five (45) days.
9.4.8 System may accept input for soil moisture warning levels.
9.4.9 System shall send notifications if moisture levels drop below user-defined floors for more than four (4) hours.

9.5 Outdoor Irrigation

The outdoor irrigation system will promote a healthy lawn as well as landscaping. The system will depend on water supplies, but will be able to automate the lawn watering process. Priority = Low.

9.5.1 Irrigation system shall be plugged into the water system.
9.5.2 System shall control any number of yard irrigation devices.
9.5.3 System shall allow user to configure any irrigation device.
9.5.4 Irrigation devices shall be configurable for type of stream, amounts of water, and turn rotation during cycle.
9.5.5 System shall test irrigation devices independently for setting configurations.
9.5.6 System shall run a test cycle of all irrigation devices simultaneously to test configuration and coverage.
9.5.7 System shall have access to any number of devices reporting soil moisture.
9.5.8 System may base water cycles on soil moisture or by a set schedule.
9.5.9 System shall allow user to set groups of irrigation devices or individual irrigation devices.
9.5.10 System shall not run irrigation devices if rain is detected.
9.5.11 System shall be able to retrieve weather forecast from the Internet.
9.5.12 System may be configured to skip a user-defined number of watering cycles if rain is in the immediate forecast (i.e., rain is 60% likely over the next 2 days).
9.5.13 System shall record the amount of water deployed through each individual irrigation device.
9.5.14 System shall present users with reports for water deployment through the lawn irrigation system.
9.5.15 System shall maintain data for lawn irrigation for no less than thirty (30) days.

9.5.16 System shall allow for motion detectors to be present in specified areas in lawn (i.e., flowerbeds or flower pots).
9.5.17 System shall allow for users to configure settings for when to activate lawn motion detectors.
9.5.18 System shall deploy counter measures (i.e., loud sound, scent repellent) when motion detectors are tripped during user-defined time periods (for scaring off animals trying to eat plants).

9.6 Outside Building Cleaning

The outdoor building cleaning system will allow automatic, periodic cleanings of all exterior surfaces of the building to promote better curb appeal. This system will depend on ready water supplies and users maintaining proper levels of chemical or cleaning agents used by the system. The system is made to be abstract enough to enable users to clean virtually any exterior surface. Priority = Medium.

9.6.1 SH shall have reservoirs for cleaning different surfaces outside the home (i.e., windows and siding).
9.6.2 System shall monitor levels of all materials needed to clean exterior surfaces.
9.6.3 System shall send notifications when materials are low.
9.6.4 System shall accept any number of cleaning devices to control.
9.6.5 System shall allow users to assign category to the type of device under the system's control.
9.6.6 System shall accept input on what type of schedule should be used to deploy devices for cleaning various exterior surfaces.
9.6.7 System shall deploy cleaning devices according to the user inputted schedule.
9.6.8 System shall store and report information about cleaning material usages on a daily basis.
9.6.9 System shall maintain cleaning material usage data for no less than thirty (30) days.

9.7 Ability to Configure Routines

The ability to configure routines will enhance the lives of the occupants, especially those portions of their lives which are routine. While life is mostly variable, there are some situations where routines are the mode of operation. All systems able to be controlled by the SH shall be presented as options to configure and set new routines within this system. It is adaptable and changeable as the lives and routines of the occupants change over time. Priority = Medium.

9.7.1 System shall allow users to configure routines.
9.7.2 System shall allow users to set alarm or wake up calls for various occupants within the house, including visitors.
9.7.3 System shall allow users to control certain activities as a result of a trigger. Example trigger-based routines would be alarm at some time, 5 minutes afterward turn on bedroom TV, 10 minutes after the alarm, turn on the shower, 15 minutes after the alarm ensure coffee maker is operational or coffee is warm.

9.8 Voice Activation

The voice activation system currently will consist of a finite set of commands to which the SH will programmatically respond. In the future this should be extended such that any commands can be programmed to control any device or system interfaced by the SH. Priority = High.

9.8.1 System shall support voice activation in major living areas (i.e., living room, kitchen, etc.).
9.8.2 System shall support commands to raise the current target temperature of the thermostat.
9.8.3 System shall support commands to lower the current target temperature of the thermostat.
9.8.4 System shall support command to draw a bath in the master bathroom
9.8.5 Master bed shall have heating element capable of warming the bed.
9.8.6 System shall support command to begin prewarming the bed in the master bedroom.
9.8.7 System shall support command to prepare the hot tub for use.
9.8.8 System shall support commands to dim or switch off lights in any room in the house.
9.8.9 System shall support commands to turn air conditioning or heating on and off.
9.8.10 System shall support commands to open windows and/or blinds on various levels of the house.
9.8.11 System shall support command to lock all points of entry.
9.8.12 System shall support command to secure the house, which would lock all points of entry and close all windows and blinds.

9.9 Driveway

The system is geared to provide ease and safety in the winter months to attempt to prevent snow accumulation on the driveway surface, and more importantly the formation of ice. Priority = Low.

9.8.1 Driveway shall have heating element installed underneath it.
9.8.2 System shall constantly monitor driveway surface temperature.
9.8.3 System will turn on driveway heating if the surface temperature of the driveway is conducive to freezing water.
9.8.4 Driveway heating element will shut off or not run if the driveway is above forty (40) degrees Fahrenheit.
9.8.5 System will monitor and record when driveway heating element is in use.
9.8.6 System may be set to only run heating surface at night, or based on time-of-day settings.

9.10 Kitchen Food Stocking

The kitchen food stocking program will provide a way for the occupants to control and view inventory from anywhere in the world. This will be helpful when shopping for groceries as well as when deciding what options may be available for dinner. Priority = Low.

9.10.1 System shall allow users to enter food associated with RFID tag into the kitchen inventory system.
9.10.2 System shall present reports to users of food inventory.
9.10.3 System shall allow users to call in (i.e., from grocery store) to check on stock of certain items in the kitchen's inventory.
9.10.4 System shall monitor and track the usage of certain items.
9.10.5 System shall present users with reports on item usage (i.e., for diets, and food spending forecasting).
9.10.6 System shall maintain item inventory and usage for no less than eighteen (18) months.
9.10.7 System shall provide interface for recipe center [ref. 9.11] to provide feedback on stock of items needed for recipe.
9.10.8 System may provide intelligent interface to create shopping list templates based on average food usage.

9.11 Kitchen Recipe Center

The kitchen recipe center will provide users with an easy way to recall and cook recipes while operating in the kitchen. The system will provide easy access to recipes and provide voice-automated help and limited automation for backing functions. The recipe center also provides some publishing mechanisms to share recipes with family and friends. Priority = Medium.

9.11.1 System shall allow users to enter recipes.
9.11.2 System shall allow users to define categories for recipes stored within the recipe center (i.e., appetizer, beef main course, dessert, etc.).

9.11.3 System shall provide touch pad interface in the kitchen for users to search, recall, and view recipes.
9.11.4 System shall provide interface for users to add, modify, and delete recipes from the repository.
9.11.5 System shall provide interface to the food stock to create grocery lists of what items may be needed for some arbitrary number of recipes.
9.11.6 System shall provide users with recipes in a specified category where all items are currently in stock. (i.e., "What can I make tonight?").
9.11.7 System shall provide users the ability to send recipes to friends electronically (i.e., email, micro Web pages, etc.).
9.11.8 System shall provide users the ability to create/print a categorized cookbook of all recipes currently within the system.
9.11.9 System shall allow users to store image file linked to any recipe within the system.
9.11.10 System shall allow user to enter assisted baking mode.
 9.11.10.1 System shall automatically preheat the oven.
 9.11.10.2 System shall verbalize order of ingredients to add.
 9.11.10.3 System shall accept verbal confirmation once item is added before instructing to add the next item.

9.12 Phone System

The phone system will be a unified approach to handling voice mail for the occupants of the house. The key functions are allowing for easier retrieval from anywhere as well as extending the system via multiple virtual inboxes. Priority = Medium.

9.12.1 System shall serve as answering machine for household.
9.12.2 System shall allow users to configure number of rings before answering.
9.12.3 System shall allow users to configure any number of phone mail boxes for recipients.
9.12.3 System shall allow users to record greeting message that will be played after user-defined number of rings.
9.12.4 System shall allow users to configure greeting message to be played for individual mail boxes.
9.12.5 System shall record messages for recipients along with date, time stamp, and incoming phone number to non-volatile memory.
9.12.6 System shall send out notification to users when they have a new message in their mailbox (i.e., email, text message, pages, etc.).
9.12.7 System shall make messages available via authenticated Web interface for user retrieval.
9.12.8 System may use voice-to-text engine to send the text representation of the message to user's email account.

9.13 Wall Pictures

The wall pictures allow for occupants of the home to allow for friends and family members to share pictures with them and have those pictures displayed on select wall monitors throughout the house. Priority = Low.

9.13.1 System shall provide wireless support for driving any number of wall-mounted monitors for picture display.
9.13.2 System shall provide Web-based interface for authenticated users to publish new photos for display on wall monitors.
9.13.3 System shall allow users to configure which pictures get displayed.
9.13.4 System shall allow users to configure which remote users can submit pictures to which wall monitor.
9.13.5 System shall support the following playback modes:
Random—display random photos.
Slideshow—display photos in order for some user-defined time.
Single—display only selected or most recently submitted photo.
9.13.6 System shall provide remote users with storage for up to 20 pictures in their repository or 100MB, whichever is greater.

9.14 Mail and Paper Notification

System to notify occupants of status and delivery events for both mail and newspaper boxes. Priority = Low.

9.14.1 System shall monitor any number of mail and newspaper boxes for motion and weight.
9.14.2 System shall allow users to set notification events for those boxes.
9.14.3 System shall send notifications when motion is detected coupled with a change in static weight of the box.
9.14.4 System shall allow user to turn off any notification events for a set period (i.e., when snow or something else may trigger the motion and weight sensors).
9.14.5 System shall permit user to query the status of any of the boxes. The status would be empty or occupied.

Glossary

agile development: class of "lightweight" software development methodologies that emphasize people over processes and feature a series of short development increments.

antipattern: a situation involving an identifiable and non-unique pattern of negative behavior. The solution to an antipattern is called a refactoring.

assertion: a relation that is true at that instant in the execution of the program.

brainstorming: requirements elicitation technique that uses informal sessions with customers and other stakeholders to generate overarching goals for the systems.

card sorting: requirements elicitation technique that has stakeholders complete a set of cards that include key information about functionality. The requirements engineer then groups the information logically or functionally.

client: see *customer.*

COCOMO: a tool that estimates software project effort and duration based on a set of parameters that characterize the project. COCOMO is an acronym for COnstructive COst MOdel.

customer: the class (consisting of one or more persons) who is commissioning the construction of the system. Also called *client* or *sponsor.*

design constraint requirement: a requirement related to standards compliance and hardware limitations.

designer as apprentice: a requirements discovery technique in which the requirements engineer "looks over the shoulder" of the customers to enable him to learn enough about the customer's work to understand their needs.

discounted payback period: the payback period determined on discounted cash flows rather than undiscounted cash flows.

divergent goals: an antipattern in which two or more major players in a situation are working against each other.

domain analysis: any general approach to assessing the "landscape" of related and competing applications to the system being designed.

domain requirements: requirements that are derived from the application domain.

Extreme Programming (XP): an agile software development methodology.

ethnographic observation: any technique in which observation of indirect and direct factors inform the work of the requirements engineer.

feature points: an extension of function points that is more suitable for embedded and real-time systems.

formal method: any technique that has a rigorous, mathematical basis.

function points: a duration and effort estimation technique based on a set of project parameters that can be obtained from the requirements specification.

goal-based approaches: any elicitation techniques in which requirements are recognized to emanate from the mission statement, through a set of goals that lead to requirements.

goal-question-metric (GQM): a technique used in the creation of metrics that can be used to test requirements satisfaction.

group work: a general term for any kind of group meetings that are used during the requirements discovery, analysis, and follow-up processes. JAD is a form of group work.

GQM: see *goal-question-metric.*

informal method: any technique that cannot be completely transliterated into a rigorous mathematical notation.

internal rate of return (IRR): a way to calculate an artificial "interest rate" for some investment for the purposes of comparing the investment with alternatives.

introspection: developing requirements based on what the requirements engineer "thinks" the customer wants.

IRR: see *internal rate of return.*

JAD: see *Joint Application Design.*

Joint Application Design (JAD): elicitation technique that involves highly structured group meetings or mini-retreats with system users, system owners, and analysts in a single venue for an extended period of time.

laddering: a technique where a requirements engineer asks the customer short prompting questions ("probes") to elicit requirements.

metrics abuse: an antipattern in which metrics are misused either through misunderstanding, incompetence, or malicious intent.

model checker: a software tool that can automatically verify that certain properties are theorems of the system specification.

negative stakeholders: stakeholders who will be adversely affected by the system.

net present value (NPV): a calculation to determine the value, in today's dollars, of some asset obtained in the future.

NFR: see *non-functional requirement.*

non-functional requirement (NFR): requirements that are imposed by the environment in which the system is to operate.

NPV: see *net present value.*

performance requirement: a static or dynamic requirement placed on the software or on human interaction with the software as a whole.

payback period: the time it takes to get the initial investment back out of the project.

PI: see *profitability index.*

protocol analysis: a process where customers and requirements engineers walk through the procedures that they are going to automate.

process clash: an antipattern in which established processes are incompatible with the goals of an enterprise.

profitability index (PI): the net present value of an investment divided by the amount invested. Used for decision making among alternative investments and activities.

prototyping: involves construction of models of the system in order to discover new features.

QFD: see *Quality Function Deployment.*

Quality Function Deployment (QFD): a technique for discovering customer requirements and defining major quality assurance points to be used throughout the production phase.

refactoring: a solution strategy to an antipattern.

repertory grids: elicitation technique that incorporates a structured ranking system for various features of the different entities in the system.

requirements engineering: the branch of engineering concerned with the real-world goals for, functions of, and constraints on systems.

return on investment (ROI): the value of an activity after all benefits have been realized.

ROI: see *return on investment.*

Scrum: an agile software methodology involving short development increments called sprints and a living backlog of requirements.

semi-formal technique: one that while appearing informal, has at least a partial formal basis.

scenarios: informal descriptions of the system in use that provide a high-level description of system operation, classes of users, and exceptional situations.

sponsor: see *customer.*

stakeholder: a broad class of individuals who have some interest (a stake) in the success (or failure) of the system in question.

task analysis: a functional decomposition of tasks to be performed by the system.

traceability: a property of a systems requirements document pertaining to the visible or traceable linkages between related elements.

UML: see *Unified Modeling Language.*

Unified Modeling Language (UML): a collection of modeling tools for object-oriented representation of software and other enterprises.

use case: a depiction of the interactions between the system and the environment around the system, in particular, human users and other systems.

use case diagram: a graphical depiction of a use case.

use case points: an effort estimation technique based on characteristics of the project's use cases.

user: a class (consisting of one or more persons) who will use the system.

user stories: short conversational text that are used for initial requirements discovery and project planning.

value engineering: the activities related to managing expectations and estimating and managing costs during all aspects of systems engineering, including requirements engineering.

viewpoints: a way to organize information from the (point of view) of different constituencies.

Wiki: a collaborative technology in which users can format and post text and images to a Web site.

workshop: a formal or informal gathering of stakeholders to determine requirements issues.

XP: shorthand notation for Extreme Programming.

Index

G

H

I

W

X

Z